World Inventory of Plutonium and Highly Enriched Uranium, 1992

Stockholm International Peace Research Institute
Pipers väg 28 S-170 73 Solna Sweden
Cable SIPRI
Telephone 46 8/655 97 00   *Telefax* 46 8/655 97 33

World Inventory of Plutonium and Highly Enriched Uranium, 1992

David Albright, Frans Berkhout and William Walker

sipri

OXFORD UNIVERSITY PRESS

1993

Oxford University Press, Walton Street, Oxford OX2 6DP
Oxford New York Toronto
Delhi Bombay Calcutta Madras Karachi
Kuala Lumpur Singapore Hong Kong Tokyo
Nairobi Dar es Salaam Cape Town
Melbourne Auckland Madrid
and associated companies in
Berlin Ibadan

Oxford is a trade mark of Oxford University Press

Published in the United States
by Oxford University Press Inc., New York

British Library Cataloguing in Publication Data
Data available

Library of Congress Cataloging in Publication Data
Data available
ISBN 0–19–829153–1

Typeset and originated by Stockholm International Peace Research Institute
Printed in Great Britain
on acid-free paper by
Biddles Ltd, Guildford and King's Lynn

Contents

Part III. Principal civil inventories

Part IV. Material inventories and production capabilities in the
threshold states

Part V. Conclusions

Appendices

Preface

The fundamental changes on the political scene, particularly the collapse of the bipolar system and the breakup of the Soviet Union, make it possible for the international community to achieve arms control agreements that were previously almost inconceivable. The years 1991–92 have brought about a genuine breakthrough in this respect. As many other problems recede with the end of the cold war, problems of controlling fissionable material are coming to the fore.

SIPRI has a long-established reputation in assembling, interpreting and publishing data related to armaments. Without detailed and accurate information it is difficult to analyse causes and effects or to have adequate debates about arms control measures.

SIPRI is therefore pleased to present the results of a research project carried out by David Albright, Frans Berkhout and William Walker—a pioneering inventory of nuclear weapon materials. Plutonium and enriched uranium are the substances that have lain at the heart of the civil and military development of nuclear technology. Amid so much concern over nuclear proliferation, and over the fate of the huge stocks of material that will soon arise from the dismantling of nuclear weapons and the reprocessing of civil fuels, this publication could not be more timely. Its comprehensiveness is one of its great merits. No country is excluded, and equal attention is given to civil and military materials. Controlling them is a problem we all have to face.

As the authors take pains to point out, much remains to be discovered about the sizes and locations of plutonium and highly enriched uranium stocks. Inventories will also change over time as more material is produced or consumed, as movements occur across international boundaries, and as the physical mediums in which the materials are held are altered. It is hoped that it will be possible to update the inventory every few years. However, I share the authors' opinion that the primary responsibility for bringing more information about plutonium and highly enriched uranium before the public lies with governments. Fifty years after the beginning of the nuclear age, it should not be the case that that there are no published international statistics on these vital materials.

Adam Daniel Rotfeld
Director of SIPRI
December 1992

Acknowledgements

This book is a result of research projects carried out over several years at the Center for Energy and Environmental Studies, Princeton University, the Federation of American Scientists, and the Friends of the Earth, USA; and at the Science Policy Research Unit, University of Sussex, UK. In the context of the nuclear weapon states, it also draws upon the work carried out by Tom Cochran, Stan Norris and their colleagues at the Natural Resources Defense Council, Washington, DC. It has been made possible by the generous financial support of The Rockefeller Brothers Fund, The Joseph Rowntree Charitable Trust and The Ploughshares Fund. The authors alone bear the responsibility for any views expressed in the book.

Like so many of our colleagues in the field of nuclear non-proliferation policy, we owe a particular debt to Hilary Palmer of The Rockefeller Brothers Fund for the great encouragement she has given us.

Many people have encouraged and helped us to compile this study. We owe a particular debt to Tom Cochran, David Fischer, Richard Guthrie, Mark Hibbs, Frank von Hippel, János Jelen, David Kyd, Alain Michel, Harald Müller, Stan Norris, Chris Paine, Jane Sharp, John Simpson, Ian Smart, Leonard Spector and Wolfgang Stoll. There are several others in government and industry that have assisted us but have asked to remain anonymous. Our special thanks to Lisa Donovan and Charlotte Huggett for their secretarial assistance, to Peter Rea for copy editing, and to Paul Claesson and Billie Bielckus at SIPRI for final editing and setting the manuscript in camera-ready format. Thanks also to Connie Wall at SIPRI for having paved the way for publication.

David Albright, Frans Berkhout and William Walker
December 1992

Glossary

Aerodynamic enrichment method	A process of uranium enrichment that is based on centrifugal effects of a fast moving uranium hexafluoride gas in very small curved-wall chambers.
Alpha particle	A charged particle emitted from the nucleus of an atom, having a mass and charge equal in magnitude to a helium nucleus.
Americium	Transuranic element with atomic number 95. Americium-241, an alpha and gamma emitter, is a decay product of plutonium-241.
Atomic bomb	A bomb whose energy comes from the fission of uranium or plutonium.
Atomic number	The number of protons in the atomic nucleus of an element.
Beryllium metal	A highly toxic steel-grey metal, possessing a low neutron absorption cross-section and high melting point, which can be used in nuclear reactors as a moderator or reflector. In nuclear weapons, beryllium surrounds the fissile material and reflects neutrons back into the nuclear reaction, considerably reducing the amount of fissile material required. Beryllium is also used in guidance systems and other parts for aircraft, missiles or space vehicles.
Beta decay	Radioactive decay involving the emission of a beta particle. This is a charged particle with the mass and charge equal to that of an electron or positron.
Blanket	A layer of fertile material, such as uranium-238 or thorium-232, placed around the core of a reactor. During operation of the reactor, additional fissile material is produced in the blanket.
Boiling water reactor	A light-water nuclear reactor in which steam is produced in the reactor and passed directly to the turbogenerator.
Burn-up	The percentage of heavy metal atoms fissioned or the thermal energy produced per mass of fuel (usually measured in Megawatt days per tonne, MWd/t).
Calutron	(*CALifornia University CycloTRON*). An electromagnetic uranium enrichment machine. Used in the Manhattan Project to produce HEU for the Hiroshima bomb and developed in the Iraqi bomb programme. Alpha machines are the first stage, producing LEU, from natural uranium; beta machines are the second stage, producing HEU from the output of the alpha machines.

CANDU

(Canadian deuterium–uranium reactor.) The most widely used type of heavy water reactor. The CANDU reactor uses natural uranium as a fuel and heavy water as a moderator and a coolant.

Cascade

A connected series of enrichment machines, material from one being passed to another for further enrichment.

Centrifuge

See ultracentrifuge.

Chain reaction

The continuing process of nuclear fissioning in which the neutrons released from a fission trigger at least one other nuclear fission. In a nuclear weapon an extremely rapid, multiplying chain reaction causes the explosive release of energy. In a reactor, the pace of the chain reaction is controlled and sustained.

Chemical enrichment

This method of uranium isotope separation depends on a slight tendency of uranium-235 and uranium-238 to concentrate in different molecules when uranium compounds are continuously brought into contact. Catalysts are used to speed up the chemical exchange.

Chemical processing

Chemical treatment of materials to separate specific usable constituents.

Cladding

The material which encases the nuclear fuel, reducing the risk of radioactive materials leaking from the fuel.

Coolant

A substance circulated through a nuclear reactor to remove or transfer heat. The most common coolants are carbon dioxide, water and heavy water.

Core

The central portion of a nuclear reactor containing the fuel elements and usually the moderator. Also the central portion of a nuclear weapon containing highly enriched uranium or plutonium.

Critical mass

The minimum mass required to sustain a chain reaction. The exact mass varies with many factors such as the particular isotope present, its concentration and chemical form, the geometrical arrangement of the material and its density. When fissile materials are compressed by high explosives in implosion-type atomic weapons, the critical mass needed for a nuclear explosion is reduced.

Depleted uranium

Uranium with a smaller percentage of uranium-235 than the 0.7 per cent found in natural uranium. It is a by-product of the uranium enrichment process, during which uranium-235 is culled from one batch of uranium, thereby depleting it, and added to another batch to increase its concentration of uranium-235.

Diversion

The deliberate removal of fissionable material in civil fuel cycles for other uses.

Draw-down	A policy of consuming stocks of nuclear material.
Dry storage	Storage of irradiated nuclear fuel in a gas (either air or an inert gas) environment.
Enrichment	The process of increasing the concentration of one isotope of a given element (in the case of uranium, increasing the concentration of uranium-235).
Facilities list	The list of nuclear facilities declared by states parties to the Treaty on the Non-Proliferation of Nuclear Weapons to the IAEA that may be subject to safeguards. In non-nuclear weapon states this includes all nuclear facilities, in nuclear weapon states it includes only facilities designated by the state.
Facility attachment	The detailed plan for applying safeguards at a particular plant. This usually defines the material balance areas, and indicates the strategic points to which the IAEA inspector may have access during inspections and at which safeguards instruments may be installed.
Fast breeder reactor	A nuclear reactor in which fuel is irradiated with high-energy neutrons and which produces more fissile material than it consumes, a process known as breeding. Fissile material is produced both in the reactor's core and through neutron capture in fertile material placed around the core (blanket).
Feed stock	Material introduced into a facility at the start of the process, such as uranium hexafluoride in an enrichment plant.
Fertile	Material composed of atoms which readily absorb neutrons to produce fissionable materials. One such element is uranium-238, which becomes plutonium-239 after it absorbs a neutron. Fertile material alone cannot sustain a chain reaction.
Fission	The process by which a neutron strikes a nucleus and splits it into fragments or 'fission products'. During the process of nuclear fission, several neutrons are emitted at high speed and radiation is released.
Fissionable material	Material, whose nuclei can be induced to fission by a neutron.
Fissile material	Material composed of atoms which fission when irradiated by slow or 'thermal' neutrons. Uranium-235 and plutonium-239 are the most common examples of fissile materials.
Fuel element	Engineered bundle of nuclear fuel pins.
Fuel pin	Single rod of basic chain-reacting material, including both fissile and fertile materials.

Fusion	The formation of a heavier nucleus from two lighter ones (usually hydrogen isotopes), with the attendant release of energy (as in a hydrogen bomb).
Gamma radiation	High-energy electromagnetic radiation emitted from nuclei as a result of nuclear reactions and decay.
Gas centrifuge process	A method of isotope separation in which heavy gaseous atoms or molecules are separated from light ones by centrifugal force and an induced counter-current flow in the swirling gas.
Gas-cooled reactor	A nuclear reactor employing a gas (usually CO_2) as a coolant, rather than water or liquid metal.
Gaseous diffusion	A method of isotope separation based on the fact that gas atoms or molecules with different masses will diffuse through a porous barrier (or membrane) at different rates. The method is used to separate uranium-235 from uranium-238. It requires large plants and significant amounts of electric power.
Gas-graphite reactor	A nuclear reactor in which a gas is the coolant and graphite is the moderator.
Graphite	One of the two elemental forms of carbon, used as a moderator in some thermal reactor types (Magnox, RBMK).
Heavy water	Water containing significantly more than the natural proportion (1 in 6500) of heavy hydrogen (deuterium) atoms to ordinary hydrogen atoms. (Hydrogen atoms have one proton, deuterium atoms have one proton and one neutron.) Heavy water is used as a moderator in some reactors because it slows down neutrons more effectively and absorbs them less (than light, or normal, water) making it possible to fission natural uranium and sustain a chain reaction.
Heavy water reactor	A reactor that uses heavy water as its moderator and natural uranium as fuel. See CANDU.
Highly enriched uranium	Uranium in which the percentage of uranium-235 nuclei has been increased from the natural level of 0.7 per cent to some level greater than 20 per cent, usually around 90 per cent.
Hot cells	Lead-shielded rooms with remote handling equipment for examining and processing radioactive materials. In particular, hot cells are used for examining spent reactor fuel.
Hydrogen bomb	A nuclear weapon that derives its energy largely from fusion. Also known as a thermonuclear bomb.
Irradiation	Exposure to a radioactive source; usually in the case of materials being placed in an operating nuclear reactor.

Isotope	Atoms having the same number of protons, but a different number of neutrons. Two isotopes of the same atom are chemically similar and are therefore difficult to separate by ordinary chemical means. Isotopes can have very different nuclear properties, however. For example, one isotope may spontaneously fission readily, while another isotope of the same atom may not fission at all. An isotope is specified by its atomic mass number (the number of protons plus neutrons) following the symbol denoting the chemical element (e.g., uranium-235 is an isotope of uranium).
Kilogram	A metric weight equivalent to 2.2 pounds.
Kiloton	The energy of a nuclear explosion that is equivalent to an explosion of 1000 tonnes of TNT.
Laser enrichment method	A still experimental process of uranium enrichment in which lasers are used to separate uranium isotopes.
Light water	Ordinary water (H_2O), as distinguished from heavy water (D_2O).
Light water reactor	A reactor that uses ordinary water as moderator and coolant and low-enriched uranium as fuel.
Light water-cooled, graphite-moderated reactor	Russian Chernobyl-type reactor cooled by water and moderated with graphite.
Low-enriched uranium	Uranium in which the percentage of uranium-235 nuclei has been increased from the natural level of 0.7 per cent up to 20 per cent, usually 3 to 5 per cent. With the increased level of fissile material, low-enriched uranium can sustain a chain reaction when immersed in light water and is used as fuel in light-water reactors.
Magnox fuel	Uranium metal fuel clad with magnesium oxide (magnox).
Magnox reactor	Gas-cooled, graphite-moderated reactor built principally in the UK and France.
Maraging steel	Special hardened steel used in the fabrication of centrifuge rotors and rocket motors.
Mass number	The number of protons and neutrons in the atomic nucleus. Elements may occur in forms displaying a range of mass numbers—i.e., plutonium-238, -239, -240, -241, -242.
Medium-enriched uranium	Uranium in which the percentage of uranium-235 nuclei has been increased from the natural level of 0.7 per cent to between 20 and 50 per cent. (Potentially usable for nuclear weapons, but very large quantities are needed.)
Megawatt	One million watts-electric (MWe): used in reference to a nuclear power plant, one million watts of electricity. One million watts-thermal (MWth): one million watts of heat.

Metric tonne	1000 kilograms. A metric weight equivalent to 2200 pounds or 1.1 tons.
Milling	A process in the uranium fuel cycle by which ore containing only a very small percentage of uranium oxide (U_3O_8) is converted into material containing a high percentage (80 per cent) of U_3O_8, often referred to as yellowcake.
Mixed-oxide fuel	Nuclear fuel containing both uranium and plutonium. Most fissions in the fuel will be of plutonium nuclei.
Moderator	A component (usually water, heavy water, or graphite) of some nuclear reactor types that slows neutrons, thereby increasing their chances of fissioning fertile material.
Natural uranium	Uranium as found in nature, containing 0.72 per cent of uranium-235, 99.27 per cent of uranium-238 and a trace of uranium-234.
Neutron	An uncharged elementary particle, with a mass slightly greater than that of a proton, found in the nucleus of every atom heavier than hydrogen. Nuclear fission is caused when a nucleus is irradiated with neutrons. Fissions may be caused by relatively low energy (thermal) neutrons, or by high energy (fast) neutrons. Fission reactors are therefore classed as either 'fast reactors' or 'thermal reactors'.
Nuclear energy	The energy liberated by a nuclear reaction (fission or fusion) or by spontaneous radioactivity.
Nuclear fuel	Basic chain-reacting material, including both fissile and fertile materials. Commonly used nuclear fuels are natural uranium and low-enriched uranium; high-enriched uranium and plutonium are used in some reactors.
Nuclear fuel cycle	The set of chemical and physical operations needed to pre-pare nuclear material for use in reactors and to dispose of or recycle the material after its removal from the reactor. Existing fuel cycles begin with uranium as the natural resource and create plutonium as a by-product. Some future fuel cycles may rely on thorium and produce the fissionable isotope uranium-233.
Nuclear fuel element	A rod, tube, plate or other mechanical shape or form into which nuclear fuel is fabricated for use in a reactor.
Nuclear fuel fabrication plant	A facility where the nuclear material (e.g., enriched or natural uranium) is fabricated into fuel elements to be inserted into a reactor.
Nuclear power plant	Any device that converts nuclear energy into useful power. In a nuclear electric power plant, heat produced by a reactor is used to produce steam to drive a turbine that in turn drives an electricity generator.

Nuclear reactor	A heat engine configured to sustain a controlled nuclear chain reaction when fuelled with fissionable materials. Reactors are of three general types: electric power reactors, plutonium production reactors and research reactors.
Nuclear waste	The radioactive by-products formed by fission and other nuclear processes in a reactor. Most nuclear waste is initially contained in spent fuel. If this material is reprocessed, new categories of waste result.
Nuclear weapons	A collective term for atomic bombs and hydrogen bombs. Weapons based on a nuclear explosion. The term is generally used throughout the text to mean atomic bombs only, unless used with reference to the nuclear weapon states (all five of which have both atomic and hydrogen weapons).
Nucleus	The part of an atom containing protons and neutrons.
On-load refuelling	Re-fuelling of nuclear reactors under power (i.e., Magnox and CANDU reactors).
Pit	The shaped core of a nuclear weapon containing fissile material, a tamper and a reflector.
Plutonium-239 (^{239}Pu)	A fissile isotope generated artificially when uranium-238, through irradiation, captures an extra neutron. It is one of the two fissile materials that have been used for the core of nuclear weapons, the other being uranium-235.
Plutonium-240 (^{240}Pu)	An isotope produced in reactors when a ^{239}Pu atom absorbs a neutron instead of fissioning. Its presence complicates the construction of nuclear explosives because of its high neutron emission and its high heat output.
Plutonium recycle	The re-use of separated plutonium as fuel in nuclear reactors.
Pond storage	Storage of irradiated fuel under water.
Power reactor	A reactor designed to produce electricity as distinguished from reactors used primarily for research or for producing radiation or fissionable materials.
Primary	The fission explosive detonated first in a thermonuclear warhead containing two or more stages (secondaries).
Production reactor	A reactor designed primarily for large-scale production of plutonium-239 by neutron irradiation of uranium-238.
Proton	A positively-charged nuclear particle, one of the two principal components of nuclei, with a mass similar to a neutron.
Radioactivity	The spontaneous disintegration of an unstable atomic nucleus resulting in the emission of sub-atomic particles.
Radioisotope	A radioactive isotope.

Reprocessing

Chemical treatment of spent nuclear fuel to separate the plutonium and uranium from unwanted radioactive waste by-products and (under present plans) from each other. Spent fuel is handled in batches known as 'campaigns'.

Research reactor

A reactor primarily designed to supply neutrons for experimental purposes. It may also be used for training, materials testing and production of radioisotopes.

Safeguards

Technical and inspection measures for verifying that nuclear materials are not being diverted from civil to other uses.

Secondary

See primary.

Separative work

A measure of the effort required in an enrichment facility to separate uranium of a given uranium-235 content into two fractions, one with a higher percentage and one with a lower percentage of uranium-235. The unit of separative work is the kilogram separative work unit (kg SWU), or separative work unit (SWU) for short. The initial material is called the 'feed'. The fraction with a higher proportion of uranium-235 is called the 'product', the other is called the 'tails'.

Significant quantity

The approximate amount of nuclear material (not just fissile material) which the IAEA considers a state would need to manufacture its first nuclear explosive. Eight kilograms of plutonium are considered significant and 25 kilograms of weapon-grade uranium are significant.

Solvent extraction

Technique for separating plutonium, uranium and fission products in a reprocessing plant using solvents.

Spent fuel

Fuel elements that have been removed from the reactor after use because they contain too little fissile and fertile material and too high a concentration of unwanted radioactive by-products to sustain reactor operation. Spent fuel is both thermally and radioactively hot.

Tails

The waste stream of an enrichment facility that contains depleted uranium. It is expressed as a percentage of the uranium-235 content and called 'tails assay.'

Thermal reactor

See neutron.

Thermal recycle

See plutonium recycle.

Thermonuclear bomb

A hydrogen bomb.

Thorium-232

A fertile material.

Tritium

The heaviest hydrogen isotope, containing one proton and two neutrons in the nucleus, produced typically by bombarding lithium-6 with neutrons. In a fission weapon, tritium is used with deuterium in a fusion process known as 'boosting' to produce excess neutrons, which set off additional fissions in the core. In this way, tritium can either reduce the amount of fissile material required, or multiply (i.e., boost) the weapon's destructive power many times. In fusion reactions, tritium and deuterium bond at very high temperatures, releasing a neutron with 14 million electron-volts of energy.

Uranium

A radioactive element with the atomic number 92 and, as found in natural ores, an average atomic weight of 238. The two principal natural isotopes are uranium-235 (0.72 per cent of natural uranium), which is fissionable, and uranium-238 (99.27 per cent of natural uranium), which is fertile.

Uranium-233 (^{233}U)

A fissionable isotope bred in fertile thorium-232. Like plutonium-239 it is theoretically an excellent material for nuclear weapons, but is not known to have been used for this purpose except in research programmes. Can be used as reactor fuel.

Uranium-235 (^{235}U)

The only naturally occurring fissionable isotope. Natural uranium contains 0.72 per cent ^{235}U; light water reactors use about 3 to 5 per cent and weapon-grade uranium has more than 90 per cent ^{235}U.

Uranium-238 (^{238}U)

A fertile material. Natural uranium is composed of approximately 99.3 per cent ^{238}U.

Uranium dioxide (UO_2)

Purified uranium. The form of natural uranium used in heavy water reactors. Also the form of uranium used to fabricate enriched uranium fuel elements.

Uranium oxide (U_3O_8)

The most common oxide of uranium found in typical ores. U_3O_8 is extracted from the ore during the milling process. The ore typically contains only 0.1 per cent U_3O_8; yellow-cake, the product of the milling process, contains about 80 per cent U_3O_8.

Uranium hexafluoride (UF_6)

A volatile compound of uranium and fluorine. UF_6 is a solid at atmospheric pressure and room temperature, but can be transformed into gas by heating. UF_6 gas (alone, or in combination with hydrogen or helium) is the feed stock in most uranium enrichment processes and is sometimes produced as an intermediate product in the process of purifying yellow-cake to produce uranium oxide.

Vessel

The part of a reactor that contains the nuclear fuel.

Weapon-grade material

Nuclear material of the type most suitable for nuclear weapons, i.e., uranium enriched to over 90 per cent ^{235}U or plutonium that is primarily ^{239}Pu.

Yellowcake

A concentrate produced during the milling process that contains about 80 per cent uranium oxide (U_3O_8). In preparation for uranium enrichment, the yellowcake is converted to uranium hexafluoride gas (UF_6). In the preparation of natural uranium reactor fuel, yellowcake is processed into purified uranium dioxide. Sometimes uranium hexafluoride is produced as an intermediate step in the purification process.

Yield
The total energy released in a nuclear explosion. It is usually expressed in equivalent tons of TNT (the quantity of TNT required to produce a corresponding amount of energy).

Zirconium
A greyish-white lustrous metal which is commonly used in an alloy form (i.e., zircalloy) to encase fuel rods in nuclear reactors.

Sources: Congressional Research Service, *Nuclear Proliferation Factbook* (US Government Printing Office: Washington, DC, 1977); Energy Research & Development Administration, *U.S. Nuclear Power Export Activities* (National Technical Information Service: Springfield, Va., 1976); Fischer, D. and Szasz, P., ed. J. Goldblat, SIPRI, *Safeguarding the Atom: a Critical Appraisal* (Taylor and Francis: London, 1985); Nero, A. V. *A Guidebook to Nuclear Reactors* (University of California Press: Berkeley, Calif., 1979); Nuclear Energy Policy Study Group, *Nuclear Power: Issues and Choices* (Ballinger: Cambridge, Mass, 1977); Office of Technology Assessment, *Nuclear Proliferation and Safeguards* (Office of Technology Assessment: Washington, DC, 1977); *Nuclear Power in an Age of Uncertainty* (Office of Technology Assessment: Washington, DC, 1984); Spector, L. S., *Nuclear Ambitions* (Westview Press: Boulder, Colo., 1990); Wohlstetter, A., *Swords from Plowshares: The Military Potential of Civilian Nuclear Energy* (The University of Chicago Press: Chicago, Ill., 1977); United Nations Association of the USA, *Nuclear Proliferation: A Citizen's Guide to Policy Choices* (UNA-USA: New York, 1983).

Abbreviations, acronyms and conventions

Acronyms and abbreviations

AGR	Advanced gas-cooled reactor
APM	Atelier Pilote Marcoule
ATR	Advanced thermal reactor
AVLIS	Atomic Vapor Laser Isotope Separation Project (USA)
BARC	Bhabha Atomic Research Centre (India)
BNFL	British Nuclear Fuels plc
BWR	Boiling water reactor
CANDU	Canadian deuterium–uranium reactor
CEA	Commissariat à l'Énergie Atomique (France)
CETEX	Army Technological Centre (Brazil)
CIA	US Central Intelligence Agency
CIS	Commonwealth of Independent States
CNEA	National Atomic Energy Commission (Argentina)
CNEN	National Nuclear Energy Commission (Brazil)
Cogema	Compagnie Générale des Matières Nucléaires
DATR	Demonstration ATR
DAE	Department of Atomic Energy (India)
DFBR	Demonstration FBR
DFR	Dounreay Demonstration Fast Reactor (UK)
DNPDE	Dounreay Nuclear Power Development Establishment
DOE	Department of Energy (USA)
DWK	Deutsche Gesellschaft für Wiederaufarbeitung von Kernbrennstoffe
EC	European Community
EdF	Électricité de France
EMIS	Electromagnetic isotope separation
ENEL	Ente Nazionale per l'Energia Elettrica
Euratom	European Atomic Energy Community
FBR	Fast breeder reactor
FBTR	Fast Breeder Test Reactor
FFTF	Fast Flux Test Facility (Hanford, USA)
GCR	Gas-cooled, graphite-moderated reactor
HEP	Head-end plant
HEU	Highly enriched uranium
HTR	High-temperature reactor
HWR	Heavy water-cooled and -moderated reactor
IAEA	International Atomic Energy Agency
INEL	Idaho National Engineering Laboratory
INF	Intermediate-range nuclear forces
IPEN	Institute of Energy and Nuclear Research (Brazil)
IPS	International Plutonium Storage
JNFS	Japan Nuclear Fuel Services Company
KfK	Kernforschungszentrum Karlsruhe

KNK	Kompakte Natriumgekühlte Kernreaktoranlage
LEU	Low-enriched uranium
LWGR	Light water-cooled, graphite-moderated reactor
LWR	Light water reactor
Magnox	Magnesium oxide
MAPI	Ministry of Atomic Power and Industry (former USSR)
MAPS	Madras Atomic Power Station (India)
MITI	Ministry of International Trade and Industry (Japan)
MOX	Mixed-oxide (uranium and plutonium)
MTR	Materials Test Reactor
MZFR	Mehrzweckforschungsreaktor
NAPS	Narora Atomic Power Station (India)
NEA	Nuclear Energy Agency, OECD
NERSA	Groupement Central Nucléaire Européen à Neutrons Rapides
NFS	Nuclear Fuel Services
NPT	Treaty on the Non-Proliferation of Nuclear Weapons
NNWS	Non-nuclear weapon state
NRDC	Natural Resources Defense Council (USA)
NWS	Nuclear weapon state
OECD	Organization for Economic Cooperation and Development
PCM	Plutonium contaminated materials
PFDF	Plutonium Fuel Development Facility
PFFF	Plutonium Fuel Fabrication Facility
PFBR	Prototype fast breeder reactor
PFPF	Plutonium Fuel Production Facility (Tokai, Japan)
PFR	Prototype Fast Reactor (UK)
PNC	Power Reactor and Nuclear Fuel Development Corporation (Japan)
PREFRE	Power Reactor Fuel Reprocessing (India)
PUREX	Plutonium uranium extraction
PWR	Pressurized water reactor
R&D	Research and development
RAPS 1	Rajasthan Atomic Power Station 1 (India)
RBMK	High-power, channel-type reactor (former USSR)
RepU	Reprocessed uranium
RERTR	Reduced Enrichment for Research and Test Reactors
RWE	Rheinische-Westfälische Elektizitätswerk AG
SAGSI	Standing Advisory Group on Safeguards Implementation (IAEA)
SAP	Service de l'Atelier Pilote
SBK	Schnell-Brüter Kernkraftwerkgesellschaft mbH
START	Strategic Arms Reduction Talks
SWU	Separative work units
THORP	Thermal Oxide Reprocessing Plant (UK)
TOP	Traitement d'Oxydes Pilote
TOR	Traitement d'Oxydes Rapides
TRR	Taiwan Research Reactor
UCOR	Uranium Enrichment Corporation (South Africa)
UKAEA	United Kingdom Atomic Energy Authority
Urenco	Uranium Enrichment Company
VAK	Versuchsatomkraftwerk Kahl
VDEW	Vereinigung Deutscher Elektrizitätswerke

VVER	Water–water power reactor
WAK	Wiederaufarbeitungsanlage Karlsruhe
WGU	Weapon-grade uranium
ZPPR	Zero Power Plutonium Reactor (Idaho, USA)

Conventions

g	Gram
GWdth	Gigawatt-days of thermal energy
GWe	Gigawatt-electric
GWe (net)	Gigawatt-electric not including power consumed by the power station itself
kg	Kilogram
kWh	Kilowatt-hour
km	Kilometre
mg	Milligram
MW	Megawatt
MWd/t	Megawatt-days per tonne of fuel
MWe	Megawatt-electric
MWth	Megawatt-thermal
MWde	Megawatt-days of electrical energy
MWdth	Megawatt-days of thermal energy
SWU	Kilogram separative work unit
t	Tonne
TWhe	Terawatt-hour of electric energy

..	Data not available or not applicable
	Nil or a negligible figure

Part I
Introduction

1. Reasons, aims and sources

I. Introduction

This volume attempts to establish the quantities of plutonium and highly enriched uranium (HEU) that existed in the early 1990s, and to identify the countries and forms in which they were held. It also presents scenarios of the amounts that may exist in the future, and considers some of the policy issues raised by these material inventories.

Plutonium and highly enriched uranium are the essential materials in nuclear weapons. Their fissioning produces the enormous amounts of energy released in atomic bombs or used to ignite thermonuclear weapons. HEU is derived by 'enriching' uranium so that it contains a much higher proportion of the fissile isotope uranium-235 (^{235}U) than is found in natural uranium. Plutonium is derived by irradiating the abundant isotope uranium-238 (^{238}U) with neutrons in a nuclear reactor, and by extracting the plutonium from its 'spent fuel' through a chemical technique known in the commercial nuclear industry as 'reprocessing'.

Plutonium and HEU were first produced in the 1940s by the USA and the USSR when they embarked on their nuclear weapon programmes. Since then, large quantities have been produced in many countries. The nuclear weapon states have acquired extensive stocks of HEU in order to manufacture warheads and to fuel submarine reactors. HEU has also been widely used, albeit in much smaller quantities, in civil research reactors. Many tonnes of plutonium have also been produced for nuclear weapons, but the largest amounts have arisen from the irradiation of uranium fuel in civil power reactors. There have long been plans to generate electricity by fuelling reactors with the plutonium extracted from these spent fuels.

II. Four security contexts

As nuclear production in both the civil and military domains expanded, and as technologies diffused, knowledge of the scale and whereabouts of plutonium and HEU inventories became increasingly important to international security in the decades following World War II. Today, there are four contexts in which this knowledge, or the lack of it, has assumed great significance. The first is that of regional nuclear proliferation—the attempts by some countries to acquire materials for their fledgling nuclear weapon programmes. Iraq's and North Korea's clandestine activities have recently demonstrated how important it is to keep track of material and technology flows and to monitor developments in the countries involved. As concern over their nuclear intentions has mounted, the international community has taken exceptional measures to establish the extent of their production capabilities and the amounts of material that they have acquired.

The second context is that of nuclear disarmament, particularly in view of the breakup of the USSR and the deep arms reductions now being undertaken by the USA and the Commonwealth of Independent States (CIS). It is expected that these developments will eventually lead to the dismantling of thousands of nuclear warheads, and the extraction from them of large amounts of plutonium and HEU. Precise material inventories will be required for managing the storage, disposal or recycling of the materials, and for providing confidence that they are well protected and will not again become available for making nuclear weapons.

The third context is that of spent fuel management, which is gaining in importance as the quantities of plutonium contained in fuels discharged from nuclear power reactors increase. While most of this plutonium is likely to remain in spent fuel assemblies, which will be stored or eventually buried, the extraction of plutonium from them is expected to increase significantly as a result of the expansion of commercial reprocessing in the United Kingdom and France in the mid-1990s, and in Japan in the early years of the next century. As the UK and France will be reprocessing fuels from other European countries and from Japan, as the plutonium will be returned to sender, and as there are plans to recycle it in power reactors, these activities, if persisted with, will result in a substantial increase in the international circulation of plutonium.

The fourth security context is the possible development of black markets in fissile materials. Unauthorized trade in plutonium or highly enriched uranium could exacerbate nuclear weapon proliferation and increase risks of nuclear terrorism. This concern reinforces the need for accurate accounting for even small quantities of fissile materials.

To the above concerns should be added the environmental and safety risks attached to radioactive nuclear materials. The accumulation of spent fuel, the extraction of plutonium and high-level wastes from it, the transport and recycling of plutonium, the dismantling of nuclear weapons—all these activities are hazardous and require strict monitoring and control.

III. The need for greater transparency

In each of these contexts, there is need for greater transparency with regard to inventories of nuclear materials. At present, knowledge of plutonium and HEU inventories is incomplete and is largely kept under wraps by governments, industrial companies and international organizations. In countries possessing nuclear weapons, or trying to acquire them, information about HEU and plutonium produced for military purposes is generally classified. In the civilian context, the position is not a great deal better. Information gathered by international agencies for safeguards purposes is held on a confidential basis and is not open to detailed public scrutiny, or even to the scrutiny of national authorities. The International Atomic Energy Agency (IAEA) and Euratom only publish broad aggregates so as to protect the identity of the countries and industrial operators providing the information. In all areas, military and civil, the information that does exist in the public domain is often inconsistent, scattered and incomprehensible to the layman.

Whereas detailed international statistics are published on oil, cotton and potatoes, for example, no equivalent records exist for these crucial nuclear materials. This report is a first attempt to bring together in one volume what is known *and not known* about the world's HEU and plutonium inventories. It seeks answers to five main sets of questions:

1. How much plutonium and HEU has been produced for military purposes in the nuclear weapon states? How much is held inside and outside nuclear weapons? How much may be released as weapons are dismantled?

2. How much plutonium is contained in spent nuclear fuels arising from nuclear power programmes? How much more may be produced over the next two decades?

3. How much plutonium has been separated from civil spent fuels? How much may be separated and recycled in the next two decades? How much HEU exists in the civil domain?

4. How much plutonium and HEU has been produced outside international safeguards in the non-nuclear weapon states? How much may have been assigned to undeclared nuclear weapon programmes?

5. What are the main gaps in information that need to be filled in order to gain a full picture of plutonium and HEU inventories?

This report is to be updated at regular intervals. We would like to believe that this will soon become unnecessary. It is time to bring an end to much of the secrecy and mystique that has surrounded these materials. It may be sensible to restrict access to information on, for instance, the precise transportation routes of fissile materials and details about storage sites in order to minimize the risks of sabotage and theft. However, most of the secrecy is a hangover from the cold war and the period when people were concerned that divulging information would put industries at a commercial disadvantage. As is discussed in chapter 13, agreement should be sought among governments to publish annually the amounts of plutonium and HEU in their possession, and to bring together the information in an international register. Establishing such a register should be placed on the agenda in the run-up to the 1995 conference to extend the Treaty on the Non-Proliferation of Nuclear Weapons (NPT).

IV. The limits to accuracy

Many hundreds of tonnes of plutonium and HEU have accumulated since the 1940s. Ideally, the amounts held in any context should be known to the nearest few kilograms (kg). Modern nuclear weapons are typically estimated to contain on average 3–4 kg of plutonium and 15 kg of weapon-grade uranium. The IAEA's Safeguards Division tries to establish inventories in non-nuclear weapon states within error margins of 8 kg of plutonium and 25 kg of HEU. These are the 'significant quantities' recommended by the IAEA's Standing Advisory Group on Safeguards Implementation (SAGSI).

These levels of accuracy are not and could not be attempted in this report. The IAEA's safeguards purpose is to detect possible diversions of what may be very small quantities of material from civil to military use. The purpose of this study is instead to provide for the first time a comprehensive empirical framework for policy analysis, and to bring together and present information in ways that make it accessible to a public audience. This is something that the IAEA would itself have difficulty doing because many of the important stocks are outside its purview, and because of the constraints on disclosure under which it operates at present.

Our estimates and the methods used to establish them have been cross-checked with people and institutions with special expertise, and much effort has gone into making the figures as accurate as possible. Despite these best efforts, significant error margins are inescapable. The figures presented here are derived mainly from knowledge of reactor histories and fuelling arrangements, and of enrichment, reprocessing and plutonium recycling programmes. Some of the statistical uncertainty is intrinsic to the subject and is shared by industries, governments and safeguards authorities. For instance, precise figures cannot be attached to the plutonium content in spent fuel unless and until it is separated during reprocessing, and there are losses of material in production processes which can usually only be estimated. However, much of the uncertainty stems from the public unavailability of information, and could be eliminated in most contexts by lifting the veil of secrecy.

Error margins vary from a few percentage points in the context of most civil programmes, to the much larger margins encountered in particular in relation to the British, Chinese, French and Soviet nuclear weapon programmes and to undeclared programmes in the threshold countries. The error margins presented in the following chapters are mainly indicative. They have not resulted from the application of formal statistical techniques, but instead suggest the degree of confidence that can be attached to the figures. Even greater uncertainty surrounds the scale, form and distribution of future plutonium and HEU inventories. Where appropriate, simple scenarios are presented, with no attempt to attach probabilities to them.

V. The scope of the book

The chief purpose of this book is to present data. Some broad policy implications are drawn in the conclusions, but analyses of consequences have elsewhere been kept to a minimum. It needs to be stressed at the outset that there is no necessary correlation between quantities of material and their political and strategic impact. In one location 50 kilograms of weapon material may give rise to greater concern than 50 tonnes in another. Stocks of material in a country that appear 'safe' today may also appear 'unsafe' tomorrow, and vice versa, as is being exemplified by current developments in the former Soviet Union and South Africa, respectively. The political significance of specific material inventories depends on the intentions and capabilities of the countries possessing them, the constraints and pressures upon them, their internal political conditions and their positions in the international order. Weighing these factors is beyond the scope of the book.

Following this introduction, chapter 2 presents a brief explanation of the main technical features of plutonium and HEU, and of their production processes. Part II is concerned with military inventories in the acknowledged nuclear weapon states. It opens in chapter 3 with an assessment of plutonium inventories, the estimates being derived mainly from information which has recently become available about military production reactors. Chapter 4 considers the extent of their stocks of weapon-grade uranium. It represents the first attempt to estimate all of these inventories using publicly available information.

Part III is concerned with the principal civil inventories. Chapter 5 assesses, by country and region, the quantities of plutonium contained in spent fuels discharged from power reactors. Chapter 6 considers the amounts of plutonium separated from these fuels, mainly in France, Japan, Russia and the UK, where civil reprocessing continues on a large scale, and offers scenarios of future plutonium arisings. The uses to which these inventories have been and may be put are examined in chapter 7. Chapter 8 briefly covers the more limited civil inventories of HEU, particularly regarding the material's use in research reactors.

Part IV is concerned with inventories in 'threshold countries' which have attempted to gain access to the materials and technologies necessary for weapon production. Chapter 9 considers those among them with the most long-standing and developed nuclear weapon programmes (Israel, India and Pakistan). Chapter 10 covers countries that have recently been the focus of international concern (Iraq, North Korea, Iran and Algeria). Chapter 11 considers countries that have recently backed away from developing nuclear weapon capabilities (Argentina, Brazil, South Africa and Taiwan).

Part V concludes the report. Chapter 12 offers an overview of plutonium and HEU inventories, drawing on the statistics presented in the preceding chapters. Chapter 13 focuses on two policy issues: the need to take steps to end the continuing poverty of public and private information on these highly significant materials; and the need to develop strategies for coping with the approaching surpluses of plutonium and HEU.

Inevitably, time and resource constraints have resulted in less attention being given to some topics than to others. One particularly regrettable gap is the management and treatment of HEU fuels discharged from submarines' propulsion reactors and from research reactors, and the subsequent recycling of enriched uranium extracted from these. The basic donkey work required to reveal how the 'back-end' of the HEU fuel-cycle has operated world-wide has yet to be carried out.

VI. Sources

The findings in this report rest on information gathered from a variety of published and unpublished sources, and from discussions with people in government, industry, the nuclear community and academia. While some of the public information is very reliable (for instance, concerning the operation of nuclear power stations), much of it is patchy and inconsistent. One of the main purposes of the report has been to organize, weed and make sense of that which is available. The reader should nevertheless keep in mind that little of the information coming from the above sources can be taken for granted, particularly where secret programmes are involved. An effort has been made throughout to explain the sources, check their reliability and to indicate where doubts remain.

Attributions to individuals can be made only in a few cases. Usually, information is imparted by persons employed by government, industry and international organizations on the agreement that they remain anonymous. Ingrained bureaucratic caution often inhibits people from talking on the record, even when the information sought has no special political or commercial value. This has not inhibited us from seeking reliable, up-to-date information. Constraints on the disclosure of sources apply especially to the chapters on the nuclear weapon states and the threshold countries. Where information has been provided in confidence, we usually indicate broadly that it has come from 'an industry official', 'an intelligence source' or 'a government official'. There are occasions where this has not been possible because to indicate the source even in these vague terms might allow the person giving the information to be identified.

It should be stressed that the authors of this book have no affiliation to any government, industrial company, security service or international organization. We are independent academic researchers, and the work has been funded by independent charitable foundations.

2. Characteristics of highly enriched uranium and plutonium and their production processes

I. Introduction

Plutonium and highly enriched uranium (HEU) have two features in common. The first is that they contain large proportions of fissile materials whose nuclei can break apart, or fission, when bombarded with neutrons, emitting more neutrons than they absorb. This gives rise to the possibility of sustained chain reactions, and thus to explosive releases of energy. The second common feature is that they are difficult and expensive to acquire. Their production requires heavy capital investment and mastery of a wide range of technologies. Were this not the case, many more countries would probably have nuclear weapons today.

In other respects, the properties of plutonium and HEU, and the nature of their production processes, are very different. While the materials are used together in nuclear weapons, they present different kinds of problem to their producers and users, and to those trying to exercise control over them. This chapter provides a brief introduction to their main characteristics.

II. Highly enriched uranium

Uranium isotopes

Isotopes are forms of an element which have nearly identical chemical and physical properties but different nuclear properties. The chemical properties of elements are fixed by the number of positively charged protons in their nuclei, and by the corresponding number of negatively charged electrons that they carry. The isotopes of an element have nuclei containing the same number of protons but different numbers of neutrons. Neutrons carry no electrical charge and can thus move with considerable freedom through atomic structures. They can penetrate an atomic nucleus and can cause the nucleus to fission, releasing a relatively large amount of energy.

Most isotopes are radioactive. The stability of an isotope is indicated by its half-life, which is the time taken for a quantity of an isotope to halve through radioactive decay. Half-lives can vary from fractions of seconds to hundreds of millions of years. Radioactive isotopes emit three main kinds of radiation when they decay: alpha-particles, which carry positive charges and consist of two protons and two neutrons (the helium-4 nucleus); beta-particles, which are energetic electrons (negatively charged) or positrons (positively charged); and gamma rays, which have no charge and are the most penetrating. Neutrons and various sub-atomic particles may also be released.

Uranium (U) has 92 electrons and 92 protons (the atomic number). Of the 14 isotopes in the sequence ^{227}U to ^{240}U (the mass numbers), ^{235}U and ^{238}U are the most important.[1] ^{235}U and ^{238}U are relatively stable isotopes with half-lives of 700 million and 4500 million years respectively. They are not strongly radioactive and can be handled by industrial workers without the need for substantial protection. In these respects, uranium contrasts with plutonium (Pu) whose principal isotopes do emit energetic radiations.

Naturally occurring uranium consists of 99.27 per cent of ^{238}U and of 0.72 per cent of ^{235}U.[2] Moreover ^{235}U, like ^{239}Pu and ^{241}Pu, fissions when irradiated with relatively low energy ('thermal') neutrons, allowing heat to be released under controlled conditions in a class of reactor called 'thermal'.[3] In thermal reactors, neutrons are slowed down or 'moderated' by materials such as graphite and water.

For nuclear weapons, and for fuel burned in most types of nuclear reactor, it is necessary to increase concentrations of ^{235}U. This is the process known as 'enrichment'. The low-enriched uranium (LEU) used to fuel commercial power reactors generally contains 2–6 per cent ^{235}U. HEU is defined as uranium containing over 20 per cent ^{235}U. For nuclear weapons, ^{235}U concentrations of 90 per cent and over are usually necessary. HEU at this level of enrichment is often referred to as 'weapon-grade uranium'. HEU with lower enrichments can be used in weapons, but the amount of material required increases substantially as enrichment levels fall.

The following five grades of uranium are commonly recognized:

1. Depleted uranium, containing less than 0.72 per cent ^{235}U.
2. Natural uranium, containing 0.72 per cent ^{235}U.
3. Low-enriched uranium, containing more than 0.72 per cent and less than 20 per cent ^{235}U.
4. Highly enriched uranium, containing more than 20 per cent ^{235}U.
5. Weapon-grade uranium, containing more than 90 per cent ^{235}U.

It should be stressed that a self-sustaining chain reaction in a nuclear weapon cannot be generated in depleted, natural and low-enriched uranium. The critical mass of uranium which can give rise to explosive releases of energy can only be constructed from materials containing high proportions of the fissile isotope

[1] The isotope ^{233}U, which is derived from neutron capture in thorium-232, has chemical properties similar to ^{235}U, but a critical mass similar to that of ^{239}Pu. It has been evaluated as a nuclear weapon material in the USA and possibly elsewhere. However, the thorium fuel-cycle has not progressed beyond the research and development (R&D) stage (it received most attention in the 1950s and 1960s), and the quantities of ^{233}U that have been produced are very small. The thorium fuel-cycle has been most actively researched in India which has large deposits of ores containing thorium.

[2] It also contains 0.006% of ^{234}U.

[3] 'Thermal' implies neutron velocities akin to the velocities of molecules in gases at room temperature (i.e., 2200 m/s).

[235]U. Thus the enriched uranium burned in conventional nuclear power reactors has no direct military value. This is recognized in the two tiers of safeguards and physical protection regulations applied to enriched uranium. Less stringent standards are applied when enrichment levels fall below 20 per cent. In contrast, a single set of regulations is applied to plutonium since it can be used in nuclear weapons in most available isotopic mixes.[4]

The corollary is that none of today's thermal power reactor designs require HEU fuels. The HEU-fuelled high-temperature reactor (HTR), prototypes of which were built in the FRG, the UK and the USA, is in abeyance. HEU is only used in submarine reactors, in a small number of breeder reactors, and in a few large research reactors. Since the mid-1970s, many research reactors around the world have been converted to operate with uranium enriched to levels below 20 per cent (see chapter 8). In France and the former USSR, submarine reactors have also been designed so that they do not require HEU.

Enrichment techniques

The techniques of gaseous diffusion and centrifuge enrichment dominate today's enrichment industry.[5] However, they are not the only techniques available. The aerodynamic techniques applied in South Africa, and the electromagnetic separation technique applied recently in Iraq, raised considerable concern as they were not subject to international safeguards. Laser and chemical techniques have also been developed in recent years. The multiplicity of technical approaches to uranium enrichment has become one of the main problems in detecting clandestine enrichment programmes in countries such as Iraq, and in tracking the technology and equipment flows connected with them.

The separative work in an enrichment plant indicates the energy expended in separating the uranium feed into enriched product and depleted uranium waste, commonly called the tails. The tails assay is the concentration of [235]U left in this waste. The unit of measurement is the kilogram separative work unit (kg SWU, usually abbreviated to SWU). The capacities of enrichment plants are expressed in SWU per year. It takes approximately 200 SWU to make 1 kg of weapon-grade uranium (uranium enriched to 90 per cent) using natural uranium feed and a tails assay of 0.3 per cent.

In brief, the following are the main enrichment processes. Details of enrichment technologies and programmes are presented in chapter 4 and appendix A.

---

[4] An exception is made for the isotope [238]Pu, which is a strong alpha-emitter and is used as a heat source in medical and space applications. Safeguards are not applied when concentrations of this isotope exceed 80%.

[5] For discussions of uranium enrichment techniques, see Krass, A. S., Boskma, P., Elzen, B. and Smit, W. A., SIPRI, *Uranium Enrichment and Nuclear Weapon Proliferation* (Taylor & Francis: London, 1983); Tait, J. H., 'Uranium enrichment', ed. W. Marshall, *Nuclear Power Technology, Vol. 2: The Fuel Cycle* (Clarendon Press: Oxford, 1983).

Gaseous diffusion

This technique exploits the property of gases whereby heavy molecules travel more slowly than light molecules. As a result, the latter will strike the walls of a vessel more frequently than the former. If parts of the vessel containing the gas are made permeable, the lighter molecules will hit the holes more often and the escaping gas will be enriched in the lighter components. Thus the uranium hexafluoride (UF_6) gas emanating at the end of a diffusion stage will be slightly enriched in the isotope ^{235}U. The final degree of enrichment attained depends on the number of stages in the cascade and on the enrichment of the initial feed.

Gaseous diffusion accounts for approximately one-half of world enrichment capacity. All past HEU production in the USA, the UK, France and China has been based on this technique. As it is an electricity-intensive process, plants are usually sited by large hydroelectric power stations (as at Oak Ridge in Tennessee) or by dedicated nuclear power stations (as at Tricastin in France).

Centrifuge enrichment

In this process the heavier molecules in a rotating gaseous mass move towards the outside of the fluid mass. It is the same technique as that used in the separation of cream in the dairy industry. In the context of uranium enrichment, it requires high precision engineering and sophisticated metallurgy because of the high speed of the centrifuges. Power consumption in a centrifuge plant ranges from 50 to 400 kilowatt-hours (KWh) per SWU, substantially less than the 2500 KWh per SWU that are typical of a gaseous diffusion plant.

The huge Soviet enrichment capacity is based on centrifuge techniques. They also form the basis of the URENCO facilities in Germany, the Netherlands and the UK. A plant to this design has also been constructed in Japan, and there are plans to build a commercial-scale facility in the USA. As is discussed in chapters 9, 10 and 11, Pakistan, Iraq and Brazil have also followed this route when seeking to attain nuclear weapon capabilities.

Aerodynamic enrichment

This technique has two variants. The Becker jet nozzle, developed in the FRG and only attempted on an industrial scale there and in Brazil (and since abandoned in both countries), exploits the mass dependence of the centrifugal force in a fast, curved flow of UF_6. The gas expands into a curved duct and the flow is split into heavier and lighter fractions by means of a skimmer. In the South African process, a mixture of hydrogen and UF_6 is allowed to swirl in a separating element which acts as a stationary-walled centrifuge. Neither aerodynamic technique has been shown to be commercially viable.

Electromagnetic separation

This technique, as recently applied in Iraq, is discussed in detail in chapter 10. In a device called a calutron, heavy and light uranium ions (atoms carrying electrical charges) follow trajectories with different curvatures in a strong magnetic field. This technique was used to produce HEU for the first US atomic weapons, but was then rejected because of its very high capital and energy intensity.

Laser enrichment

There has been speculation since the early 1970s that laser enrichment will provide the basis of the next generation of enrichment plants. If made to operate effectively, it would have the lowest power consumption and the greatest mechanical simplicity among the physical techniques, although it would still be a complex and difficult process. The isotopes of uranium can be selectively excited by high-energy lasers. There are two technical approaches. In the atomic route, ^{235}U is selectively excited using tuneable lasers, and the resulting ionized atoms are separated electromagnetically. In the molecular route, selective infra-red absorption of $^{235}UF_6$ gas is followed by further irradiation at infra-red or ultraviolet frequencies, allowing the dissociation of the excited molecules or their chemical separation.

The first of the above techniques was adopted in the USA's Atomic Vapor Laser Isotope Separation (AVLIS) project. In 1985, the US Department of Energy (DOE) chose the AVLIS process when planning new enrichment capacity for the 1990s and beyond. However, the construction of AVLIS plants has been delayed owing to development problems and the mounting oversupply in the world enrichment market.

Plasma separation and chemical exchange

Two other techniques are still at the research and development (R&D) stage. The first is the plasma separation process, in one version of which uranium atoms are exposed to low-energy radio frequency waves resonating with the 'cyclotron frequency' of ^{235}U ions. By rotating these ions, they can be collected on electrically charged plates. The second and more important technique is the chemical exchange process. This method depends on a slight tendency of ^{235}U and ^{238}U to concentrate in different molecules when uranium compounds are continuously brought into contact. Catalysts are used to speed up the chemical exchange. Pilot plants using this technique have been constructed in France and Japan.

HEU recycling

For many years, HEU from spent submarine and research reactor fuels has been recovered by chemical reprocessing.[6] It has either been used as a fuel for plu-

---

[6] HEU extracted from spent fuels can only be re-enriched with difficulty, particularly because of the presence of the isotopes ^{232}U and ^{236}U. The ^{232}U increases radiation risks, while ^{236}U can form bubbles in the middle of the cascade, impeding flows of the other isotopes.

tonium production reactors, as in the cases of domestic and foreign spent HEU fuels reprocessed at the Savannah River and Idaho National Laboratory facilities in the USA; or it has been blended with depleted uranium, to produce low-enriched uranium which can be used in power reactors. The latter option has been followed by the USSR in providing uranium fuel for RBMK (high-power channel-type) reactors.

What is now in prospect is that large amounts of HEU, mostly containing more than 90 per cent of ^{235}U, will be released from dismantled US and CIS nuclear weapons (see chapters 4 and 12). If diluted with depleted uranium, it would give rise to substantial quantities of low-enriched uranium, which could be used to fuel power reactors. A deal has recently been struck between the Russian and US governments whereby a substantial proportion of the former Soviet Union's stockpile of HEU will be purchased by the USA. After dilution, it will be introduced into the civil fuel cycle.

The point to be stressed here is that recycling HEU is technically straight-forward. There are no technical obstacles to reducing stocks of this weapon material. Moreover, there is a ready commercial market for the resulting uranium fuel. If there is a problem, it is that introducing HEU to the world market when there is already a surfeit of enrichment supply could depress the prices of natural and enriched uranium, or at least keep prices at their present low levels. While the uranium industry might lose, the electricity consumer would benefit from the low prices, although utilities would have to guard against the risk that under-investment in uranium supply would lead to future scarcity and price inflation.

As seen below, this is not the case for plutonium. Plutonium stocks cannot be extinguished in this way, and there will be strong economic disincentives to recycling it in power reactors unless uranium prices rise very substantially.

III. Plutonium

Unlike uranium, plutonium is entirely a manufactured material. Several of plutonium's isotopes are also highly radioactive, and its processing into weapon components or into fuel is far from straightforward. Plutonium is therefore a difficult and hazardous material to work with. In a number of respects, it gives rise to greater technical, industrial and regulatory complexity than uranium.

Plutonium isotopes and grades

Plutonium-239 is produced in a nuclear reactor by bombarding ^{238}U with neutrons. Neutron capture turns ^{238}U into ^{239}U, which decays via neptunium-239 in a matter of days to ^{239}Pu. While ^{239}Pu has a half-life of 24 000 years and is thus a relatively stable isotope, it is readily fissioned by both thermal and fast neutrons. It also absorbs neutrons, in addition to being fissioned by them, resulting in the formation of the isotope ^{240}Pu. Subsequent neutron captures lead to accumulations of ^{241}Pu, ^{242}Pu and ^{243}Pu.

Plutonium-239 and ^{241}Pu are more susceptible to fissioning than the other plutonium isotopes, and are alone in being fissionable by thermal neutrons. They are therefore usually referred to as the 'fissile' isotopes of plutonium. In plutonium commerce, quantities are often expressed in terms of the amount of these fissile isotopes in a particular batch of material. The combined weight of ^{239}Pu and ^{241}Pu is then recorded as Pu_{fiss}, as distinct from Pu_{tot} which refers to the total weight of plutonium isotopes in the batch.

While the plutonium used in nuclear weapons usually contains very small quantities of ^{241}Pu, this is not the case with the plutonium derived from most power reactor fuels. ^{241}Pu can cause serious problems in plutonium handling since it decays to americium-241 (^{241}Am), which is an intense X-ray and gamma-emitter. ^{241}Pu has a half-life of 13.2 years so that substantial quantities of ^{241}Am can quickly accumulate in plutonium separated from reactor fuels, leading to the need for heavier shielding to protect workers involved in the handling and safeguarding of plutonium. More recently, limits have also been imposed on the acceptable amount of spontaneous neutron radiation from even-numbered isotopes. As a rule of thumb, plutonium derived from power reactor fuels has to be recycled within three years of separation. If left longer, the costs of fuel fabrication rise steeply because of the need to provide extra protection or to extract chemically the americium before fuel fabrication.

While ^{241}Pu is a problem in nuclear commerce, the even-numbered isotopes ^{240}Pu and ^{242}Pu are irritants for nuclear weapon designers.[7] They fission spontaneously, producing energetic neutrons which can result in premature initiation of a chain reaction in the plutonium contained in a warhead. As a result, the weapon can 'fizzle', reducing its explosive yield. There are ways of avoiding this problem, even with plutonium containing substantial fractions of the even-numbered isotopes, without major compromises in yield, reliability, weight or efficiency. The heat emitted by the even-numbered isotopes is a further complication for the weapon designer. The preference is always for material with high concentrations of ^{239}Pu.

A distinction is therefore commonly made between different grades of plutonium. The following definitions laid down by the US Department of Energy have gained wide currency:

1. Weapon-grade plutonium, containing less than 7 per cent ^{240}Pu.
2. Fuel-grade plutonium, containing from 7 to 18 per cent ^{240}Pu.
3. Reactor-grade plutonium, containing over 18 per cent ^{240}Pu.

'Super-grade plutonium' is also sometimes used to denote plutonium containing between 2 and 3 per cent ^{240}Pu. The term 'weapon-usable plutonium' has no precise definition. It has been adopted on occasion to convey the mes-

[7] For a discussion of weapon design and grades of plutonium, see Mark, J. C., *Reactor-Grade Plutonium's Explosive Properties* (Nuclear Control Institute: Washington, DC, Aug. 1990).

sage that most isotopic mixtures of plutonium can be used in nuclear weapons, or to imply that a given quantity of plutonium is in separated form.

Plutonium production: irradiation of reactor fuels

The production of plutonium is carried out in two main industrial stages. The first involves the irradiation of uranium fuels in nuclear reactors. The second involves the chemical separation of plutonium from the uranium, transuranic elements and fission products contained in discharges of irradiated fuel. The second technique is known as 'reprocessing' when applied commercially. The need to construct both reactors and reprocessing facilities to acquire plutonium is one reason why some aspiring nuclear weapon states have favoured the use of HEU in recent years.

Although they can overlap, plutonium is produced in two different contexts. In the military context, the reason for irradiating nuclear fuel is to acquire stocks of weapon-grade material for use in nuclear warheads: plutonium supply is the *raison d'être*. In the civilian context, the purpose is to generate electricity, plutonium being a by-product which may or may not have further uses. The days have gone when it was considered imperative to accumulate plutonium as the starting fuel for a coming generation of advanced reactors. The isotopic content of discharged plutonium is a serious concern of nuclear weapon designers, but is less important to electricity producers.

As distinct from the production of uranium with high concentrations of ^{235}U, the technique of isotopic enrichment has not been used to produce weapon-grade plutonium from lower-grade material.[8] R&D programmes were launched in the 1980s to develop laser techniques for enriching plutonium, and plans were hatched, especially in the USA, to extract weapon-grade plutonium by these means. These plans came to nought because of the reduced demand for weapon-grade plutonium as weapon programmes were curtailed. However, it is recognized that the successful development of plutonium enrichment techniques would ease access to weapon-grade material and thus simplify the acquisition of high-performance weapons. Countries with nuclear power programmes could, in principle, obtain weapon-grade plutonium from their highly irradiated power reactor spent fuel if plutonium enrichment became a viable option.

Instead, the nuclear weapon producers have achieved the desired isotopic content of plutonium mainly by controlling the extent to which uranium fuel elements are irradiated in nuclear reactors. This is known as the fuel burn-up, whose unit of measurement is megawatt-days per tonne of uranium fuel (MWd/t). Weapon-grade plutonium is produced by operating reactors at low burn-ups—usually below 1000 MWd/t—so that insufficient time elapses for a

---

[8] An indirect but very expensive way of achieving the same end is to bombard a blanket of ^{238}U in a fast reactor. The reactor can be fuelled with low-grade plutonium, while weapon-grade material can be extracted from the blanket. Among the nuclear weapon states, this has only been practised by France and perhaps the former Soviet Union (see chapter 3).

Table 2.1. Plutonium half-lives, and weapon-grade and reactor-grade isotopic concentrations at fuel discharge

Isotope	Half-life (years)	Weapon-grade isotopic concentrations (typical, %)	Reactor-grade isotopic concentrations (typical, %)		
			PWR[a] (33 000 MWd/t[d])	Magnox[b] (5000 MWd/t)	CANDU[c] (7500 MWd/t)
^{238}Pu	86.4	..	1.3	..	..
^{239}Pu	2.4 x 10^4	93.0	56.6	68.5	66.6
^{240}Pu	6.6 x 10^3	6.5	23.2	25.0	26.6
^{241}Pu	13.2	0.5	13.9	5.3	5.3
^{242}Pu	3.8 x 10^5	..	4.7	1.2	1.5

[a] Pressurized water reactor.
[b] Magnesium oxide reactor.
[c] Canadian deuterium–uranium reactor.
[d] Megawatt-days per tonne of uranium fuel.

Sources: Organization for Economic Co-operation and Development, Nuclear Energy Agency, *Plutonium Fuel: an Assessment* (OECD: Paris, 1989), tables 2 and 4; authors' data.

substantial buildup of ^{240}Pu. To achieve this, irradiated fuel is usually discharged a few weeks after insertion, the precise time depending on the type of reactor.

The military fuel-cycle is discussed in chapter 3. Civil power reactors are operated at higher burn-ups in order to optimize the energy output from a given amount of fissile material. Power reactors fuelled with natural uranium, such as the gas-cooled Magnox (magnesium oxide) reactor developed in the UK and France, and the Canadian deuterium–uranium (CANDU) reactor, have burn-ups in the range 3000–8000 MWd/t. The most common type of thermal power reactor, the pressurized water reactor (PWR) which is fuelled with enriched uranium, is typically operated at 30 000–40 000 MWd/t. As table 2.1 shows, the concentrations of the even-numbered isotopes become substantial at these burn-ups. It should also be noted that the concentrations of total fissile plutonium (^{239}Pu plus ^{241}Pu) are not dissimilar for these reactor types, but that PWR fuel contains relatively low concentrations of ^{239}Pu and high concentrations of the troublesome ^{241}Pu.

The trend is towards still higher burn-ups, with many utilities aiming for 50 000–60 000 MWd/t for PWR fuel by the end of the century. As the irradiation period is extended, the energy extracted from the fissioning of ^{239}Pu and ^{241}Pu increases as that from ^{235}U decreases. In effect, a high burn-up strategy is a cheap substitute for recycling plutonium from lower burn-up fuels. These high burn-up spent fuels will also have increasing concentrations of the isotopes ^{238}Pu, ^{240}Pu, ^{241}Pu and ^{242}Pu, which can have detrimental consequences for the economics of plutonium recycling.

Plutonium production: spent fuel reprocessing

In the military fuel cycle, all plutonium is routinely separated from the irradiated fuels discharged from production reactors. In contrast, most of the spent fuel emanating from civil power reactors is today held in store. By the end of the century, one-fifth or less of world spent fuel arisings will have been reprocessed. Nevertheless, as shown in chapter 6, the reprocessing of spent fuels particularly at facilities in France, Russia and the UK may still give rise to very large amounts of separated plutonium.

In contrast to enrichment, only one process is currently used to extract plutonium from spent reactor fuels.[9] This is the PUREX process developed in the USA in the late 1940s and early 1950s. Plutonium separation occurs in three main stages. In the first, the spent fuel assemblies are dismantled and the cladding around the fuel elements is removed by physical or chemical means. In the second stage, the extracted fuel is dissolved in hot nitric acid. In the third and most complex stage, the plutonium and uranium are separated from other actinides and fission products, and then from each other, by a technique known as 'solvent extraction'. Tributyl phosphate is commonly used as the organic solvent in a kerosene-type diluent in the PUREX process. The plutonium and uranium are usually taken through several solvent extraction cycles to reach the required levels of purity.

In modern reprocessing plants, less than 1 per cent of the plutonium contained in spent fuel may end up in the wastes for final disposal. This high extraction efficiency means that there is a close relationship between capacity utilization and plutonium output. In assessing the quantities of plutonium separated, the most important variables are the characteristics of the spent fuel inputs—their types and irradiation histories—and the strategies adopted for reprocessing different batches of fuel. The separation process itself is not a major source of uncertainty, at least at the levels of accuracy attempted in this book.[10] Reprocessing also has no effect on isotopic concentrations of plutonium; in this respect, outputs are identical to inputs.

Plutonium recycling

Fissile plutonium can be used as an alternative to fissile uranium in reactor fuel. In the 1970s, the future expansion of nuclear power seemed predicated upon the recycling of plutonium, since it was believed that reserves of low-cost uranium would soon be depleted at the growth rates of nuclear electricity production then envisaged. Today, however, uranium is cheap and abundant, and the stock

[9] The chemical properties of plutonium are discussed in Cleveland, J. M., *The Chemistry of Plutonium*, American Physical Society, La Grange Park, Ill., 1979.. A general discussion of reprocessing technology can be found in Allardice, R. H., Harris, D. W. and Mills, A. L., 'Nuclear fuel reprocessing in the UK', ed. W. Marshall (note 5).

[10] Measuring material balances in bulk reprocessing plants at the kilogram accuracies required by safeguards authorities is, however, a serious problem. A good general discussion of safeguarding reprocessing plants is provided by Lovett, J. E., IAEA Department of Safeguards, 'Nuclear material safeguards for reprocessing', document no. STR-151/152 (International Atomic Energy Agency: Vienna, 1987).

Table 2.2. Neutron cross-sections

Isotope	Thermal neutrons			Fast neutrons		
	Fission cross-section (barns)[a]	Capture cross-section (barns)	Fission/total cross-section (%)	Fission cross-section (barns)	Capture cross-section (barns)	Fission/total cross-section (%)
^{235}U	579	100	85	2.0	0.5	80
^{238}U	..	3	–	0.05	0.3	17
^{239}Pu	741	267	74	1.9	0.6	76
^{240}Pu	..	290	–	0.4	0.6	40
^{241}Pu	1 009	368	73	2.6	0.6	81
^{242}Pu	..	19	–	0.3	0.4	43
^{241}Am	3	832	0.4	0.4	1.9	17

[a] A barn is the unit of effective cross-sectional area of the nucleus equal to 10^{-28} m^2.

Source: Farmer, A. A., 'Recycling of fuel', ed. W. Marshall, *Nuclear Power Technology, Vol. 2: Fuel Cycle* (Clarendon Press: Oxford, 1983), tables 9.1 and 9.2; authors' calculations.

of nuclear power stations has stopped increasing. The real price of uranium is not expected to rise substantially for many years or even decades, while the economics of plutonium recycling look increasingly unfavourable (see chapter 7).

In terms of nuclear physics, plutonium is more suitable for recycling in fast reactors than in thermal reactors. The fission and capture cross-sections of an isotope are the technical terms used to indicate the probability of neutron absorption by an atomic nucleus. As they imply, the former indicates the probability that absorbed neutrons will fission nuclei, while the latter indicates the probability that neutrons will be captured without fissions occurring. Whereas the fission cross-sections of ^{239}Pu and ^{241}Pu irradiated with thermal neutrons are slightly higher than those of ^{235}U, this uranium isotope is the better fuel because the isotopes of plutonium have considerably higher capture cross-sections (see table 2.2). In a thermal reactor, around 85 per cent of the neutrons absorbed by ^{235}U therefore cause fissions (the remainder are captured to produce ^{236}U), while the proportion for ^{239}Pu is 74 per cent. Moreover, ^{240}Pu and ^{241}Pu also have high capture cross-sections, as does ^{241}Am. As a result, a larger amount of fissile plutonium than fissile uranium is required for a given energy output in thermal reactors.

Plutonium is typically not used on its own to fuel reactors. It is blended with natural or depleted uranium in so-called mixed-oxide (MOX) fuels. The costs of fabricating MOX fuels for thermal reactors are today five or six times higher than the costs of fabricating ordinary uranium fuels (see chapter 7), so that the prices of MOX fuel are higher even when the plutonium used in them is regarded as a free good. Utilities wishing to raise the burn-up of new fuels like MOX may also face licensing delays, causing burn-ups to lag behind those allowed with uranium fuels. As a result, MOX fuels are unlikely to be used at burn-ups much above 30 000 MWd/t for several years to come. For all these reasons, MOX fuels will be generally uncompetitive with uranium fuels when used in thermal reactors, unless uranium prices rise steeply and offset the higher manufacturing costs.

In fast reactors, plutonium does not suffer from the same physical disadvantages as uranium fuel. Around 76 per cent of the neutrons absorbed by ^{239}Pu result in fissions, compared with 80 per cent in the case of ^{235}U. In fast reactors, higher concentrations of ^{235}U or ^{239}Pu—typically in the range of 15–25 per cent, compared to 3–6 per cent with light water reactor (LWR) fuel—are required. While neutrons are more energetic, table 2.2 shows that the fission and capture cross-sections of the fissile isotopes are much lower than in thermal reactors. Moreover, the fission cross-sections of the isotopes ^{238}U, ^{240}Pu and ^{242}Pu are relatively high, as a proportion of the total cross-sections, when materials are being irradiated with fast rather than thermal neutrons, although they are still low compared to the cross-sections exhibited by the fissile isotopes. Higher enrichment levels are required to compensate for the lower cross-sections in fast reactors, and to prevent the average rate of fissioning being dragged down by these other isotopes.

Although fast reactor cores therefore require large initial inventories of plutonium or HEU for this reason, fast reactors also seemed attractive because they could 'breed' plutonium. Once running, it was expected that they would produce more fissile fuel than they consumed, mainly through ^{238}U neutron capture in uranium 'blankets' placed adjacent to the reactor core.

Despite these advantages, the prospects for fast reactors have diminished greatly in recent years. High capital costs, operational difficulties and doubts over safety have led all countries with fast reactor programmes to revise their plans. No country now claims that it will construct fast reactors in significant numbers before the middle decades of the next century.

The heavy demand for plutonium to fuel fast reactors that was forecast in the 1970s has therefore largely evaporated. As shown in chapter 7, utilities in Europe and Japan have turned to MOX recycling in thermal reactors as a means of consuming the large quantities of plutonium that they will soon acquire as their spent fuels are reprocessed in France and the UK. Particularly in non-nuclear weapon states, utilities and governments fear the political consequences at home and abroad of amassing large stocks of separated plutonium.

Plutonium from dismantled US and Soviet weapons might also become available for recycling in the 1990s (see chapters 3 and 12). As it is weapon-grade material, it is easier to handle than the plutonium from reactor spent fuels which contain significant concentrations of the higher isotopes, and in particular of the ^{241}Pu which gives rise to the gamma-emitting ^{241}Am. However, the USA is opposed to plutonium recycling, and it is questionable whether Russia will have the technical or financial resources to use its plutonium stock in this way.

In chapters 3 and 7 it is shown that the amounts of plutonium emerging from weapon dismantlement and civil reprocessing over the next two decades, and that will end up being recycled, are very uncertain. With strong economic dis-incentives to plutonium recycling, the majority of plutonium arisings from these sources might have to be stored and eventually treated as wastes. It should be noted, however, that plutonium storage also has its problems. Separated pluto-nium requires extensive physical protection and safeguarding, and as a toxic material needs special precautions so that it does not endanger health.

In technical and environmental terms, dealing with surplus plutonium is thus a more difficult problem than reducing stocks of HEU. The latter will neverthe-less have the greatest commercial impact. The quantities of HEU are much greater than those of plutonium, and the HEU will be entering a market for enriched uranium that already suffers from over-supply. These policy dilemmas are discussed in further detail in chapter 13.

Part II
Military inventories in the nuclear weapon states

Part II covers the military nuclear programmes of the five acknowledged nuclear weapon states (the USA, Russia, the UK, France and China). Article IX.3 of the 1968 Treaty on the Non-Proliferation of Nuclear Weapons (NPT) states that 'a nuclear weapon state is one which has manufactured and exploded a nuclear weapon or other nuclear explosive device prior to January 1, 1967'. (India and Israel, which might qualify for inclusion under this definition if a later date were attached, are discussed in part IV, chapter 9.)

Early in 1992 Russia was accepted as the USSR's successor state, and thus as a nuclear weapon state (NWS) under the Treaty. As nuclear weapons located in the other republics of the former Soviet Union are scheduled to be withdrawn to Russia, where all the significant nuclear production facilities are also located, all the former Soviet Union's military inventories of plutonium and HEU are expected to end up in Russia. However, this will take time, and at the time of writing it is unclear where title to the materials from weapons once located in the Ukraine, Kazakhstan and Belarus will eventually reside. To avoid confusion, this book assesses the inventories located on the combined territories of the members of the Commonwealth of Independent States (CIS)—that is, the former Soviet Union less the Baltic states.

In each NWS, the amounts of plutonium and HEU contained in nuclear warheads, or produced for their manufacture, remain classified information. During the cold war, secrecy about plutonium production and usage stemmed from the perceived need to protect information which could be used to derive estimates of operational weapon capabilities. It also suited governments to keep their plutonium production activities out of the public eye, in order to minimize the constraints placed upon them. Now that the cold war is over, and the production of special nuclear materials for weapon programmes has largely ceased in these countries, it must be hoped that the veils of secrecy will soon be lifted.

Until this happens, there will be much less public information about military than about civilian plutonium inventories. Nevertheless, the methods and information required to describe them have gradually been established over the past decade, particularly through research programmes conducted at Princeton University and the Natural Resources Defense Council (NRDC) in the United States, and through the work of individual researchers such as Simpson in Britain, and Pharabod, Genestout and Lenoir in France. Although error margins remain very large, particularly with regard to HEU, estimates can now be attempted.

Chapter 3 is concerned with plutonium inventories, and Chapter 4 with HEU inventories. (Estimates of the amounts of material that may be extracted from dismantled warheads are presented in part V, chapter 12).

3. Inventories of military plutonium in the nuclear weapon states

I. Introduction

This chapter assesses the amounts of plutonium produced for military purposes by the USA, the UK, France, China and the former Soviet Union.[1] It opens with a brief description of how the plutonium for nuclear warheads is acquired.

II. The production process

The production of plutonium for warheads typically begins with the irradiation of uranium in 'production reactors' (see figure 3.1). In most cases these are specifically designed and operated to produce plutonium with a high concentration of the fissile isotope ^{239}Pu and a low concentration of other isotopes, notably ^{240}Pu, a strong neutron-emitter the presence of which complicates the design and manufacture of nuclear weapons. Weapon-grade plutonium is usually defined as plutonium containing less than 7 per cent of ^{240}Pu.

Because natural uranium was more readily available than enriched uranium in the early stages of their weapon programmes, the nuclear weapon states (NWS) developed three types of reactor that utilize natural uranium fuel. The first was the light-water-cooled, graphite-moderated reactor (LWGR), which was the design commonly used in the USA, the USSR and China (see table 3.1). The second was the gas-cooled, graphite-moderated reactor (the GCR, or Magnox reactor), which became the staple of the British and French weapon programmes. The third was the heavy-water-cooled and -moderated reactor (HWR), units of which were constructed by the USA and USSR, partly to serve as dual-purpose plutonium and tritium producers.[2]

---

[1] Important texts on this subject are: von Hippel, F., Albright, D. H. and Levi, B. G., *Quantities of Fissile Materials in US and Soviet Weapons Arsenals*, PU/CEES Report no. 168 (Center for Energy and Environmental Studies, Princeton University: Princeton, N. J., July 1986); Cochran, T. B., Arkin, W. M., Norris, R. S. and Hoenig, M. M., *Nuclear Weapons Databook, Vol. II: US Nuclear Warhead Production* (Ballinger: Cambridge, Mass., 1987); Cochran, T. B. and Norris, R. S., *Soviet Nuclear Warhead Production* (Natural Resources Defense Council: Washington, DC, Feb. 1991); Simpson, J., *The Independent Nuclear State: the United States, Britain and the Military Atom* (MacMillan: London, 1986); Genestout, M. and Lenoir, L., 'Quelques vérités (pas toujours bonnes à dire) sur les surgénérateurs', *Science et Vie*, vol. 131, no. 781 (Oct. 1982), pp. 16–25 and 164–68; Pharabod, J.-P., *La production de plutonium pour le programme d'armement Français et le role du surgénérateur Superphénix* (Laboratoire de Physique Nucléaire des Hautes Energies, Ecole Polytechnique: Paris, June 1986).

[2] Tritium, the isotope hydrogen-3, is used to 'boost' the yield of the primary stages of nuclear warheads through a fusion reaction which produces copious amounts of high-energy neutrons that fission the plutonium or HEU more efficiently. It is produced mainly by irradiating lithium-6 in production reactors.

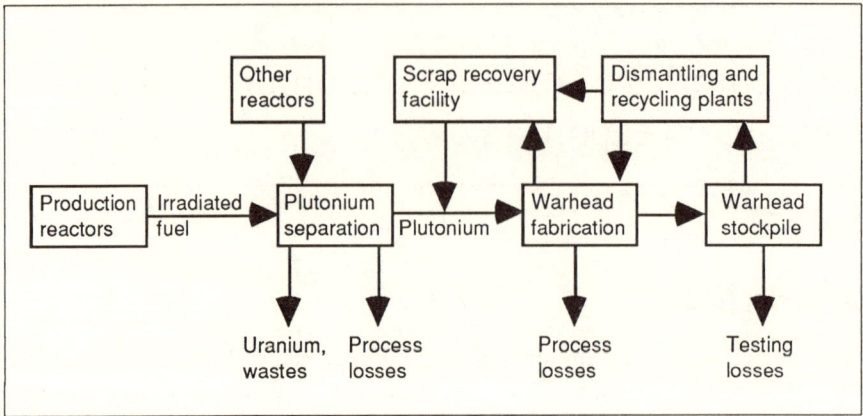

Figure 3.1. Plutonium in warhead production

To avoid the buildup of the undesirable ^{240}Pu and of higher isotopes of pluto-nium, uranium fuel is discharged from production reactors after a shorter period than from commercial reactors. Typically, fuel elements will be irradiated for a few weeks, involving burn-ups of hundreds of megawatt-days per tonne (MWd/t), compared to the tens of thousands of MWd/t in commercial opera-tion. A consequence of these low burn-ups is that the discharged fuel is less radioactive, and does not have to be left for so long to cool before processing. The time which elapses between the insertion of fuel in a reactor and the extraction of plutonium can thus be considerably shorter in the military than in the civil fuel cycle.

As figure 3.1 shows, the plutonium is extracted from irradiated fuel in pluto-nium separation plants, using chemical processes that either pre-date or are identical to those applied commercially.[3] In the NWS, plutonium separation often occurs on the sites of production reactors (see table 3.1). After conversion into suitable metallic form at reprocessing or weapon fabrication sites, the plu-tonium is cast, machined and assembled into the 'pits' which form the fissile devices used either alone or in the first stages, or primaries, of thermonuclear weapons.

In most NWS the amount of plutonium extracted for weapon purposes has been considerably less than might be expected from a count of the numbers of fabricated warheads. In the USA, for instance, the quantity of plutonium that has been produced appears to have been adequate for far less than the 60 000 warheads that have been manufactured in the USA over the past 45 years. The reason for this is that material has routinely been taken from old weapons and

---

[3] The term 'plutonium separation' and 'reprocessing' are used in the military and civil contexts, respectively, to indicate the same process.

Table 3.1. Historical sources of weapon-grade plutonium

Country	Main production site/location	Reactor/plant number, type[a] (designation)	Other sources of weapon-grade plutonium
USA	Hanford Reservation, Washington	9 LWGR (B, D, F, DR, H, C, KE, KW and N); PUREX reprocessing plant	Original Chicago and Clinton (Oak Ridge, Tenn.) piles; material imported from the UK
	Savannah River, North Carolina	5 HWR (R,P, L, K and C); F and H reprocessing plants	
USSR	Kyshtym complex, Chelyabinsk, Urals	5 LGWR: A, IR, AV-1, AV-2 and AV-3; 1 HWR; Mayak reprocessing plant	Original piles at Kurchatov Research Institute, Moscow
	Siberian Chemical Combine, Tomsk, Siberia	5 LWGR; unnamed reprocessing plant	
	Dodonovo, Krasnoyarsk, Siberia	3 LWGR; unnamed reprocessing plant	
UK	Sellafield (Windscale), Cumbria	8 GCR (Calder Hall and Chapelcross (off-site)); B204 and B205 reprocessing plants	Original Windscale piles; initial discharges from civil Magnox reactors
France	Marcoule, Côtes du Rhône	3 GCR (G1, G2 and G3); 2 HWR (Celestin 1 and 2); UP1 reprocessing plant	Civil Magnox reactors (esp. Chinon); uranium blankets in Rapsodie and Phénix fast reactors
	La Hague, Normandie	UP2 reprocessing plant	
China	Jiuquan complex, Subei County	1 LWGR; 1 unnamed reprocessing plant	
	Guangyuan, Sichuan	1 LWGR;1 unnamed reprocessing plant	

[a] GCR: gas-cooled, graphite-moderated reactor (magnox reactor); HWR: heavy-water-cooled and -moderated reactor; LWGR: light-water-cooled, graphite-moderated reactor.

Source: Cochran, T. B., Arkin, W. M., Norris, R. S. and Hoenig, M. M., *Nuclear Weapons Databook, Vol. II: US Nuclear Warhead Production* (Ballinger: Cambridge, Mass., 1987); Cochran, T. B. and Norris, R. S., *Soviet Nuclear Warhead Production* (Natural Resources Defense Council: Washington, DC, Feb. 1991); authors' data.

recycled. This has been achieved by dissolving the plutonium pits and extracting the material by standard chemical methods.[4]

In principle, all material from retired warheads can be recycled in this way. In the UK and the USA, where production of weapon-grade plutonium was cut back in the mid- to late-1960s, there has subsequently been heavy reliance on

[4] This is carried out either in specialized facilities at the sites where warhead plutonium is processed, or by introducing batches of nitrate solution containing warhead material into the appropriate parts of plutonium separation plants.

recycled material when constructing new warheads. The USSR (and probably also France) relied less on recycled weapon material.[5] Instead, it was customary for warheads from old weapons either to be reallocated to new delivery vehicles, or placed in store. It is not known whether warhead materials have been recycled in China.

The NWS have not only derived weapon-grade plutonium by producing it directly in production reactors, or by recycling material from retired warheads. The USA in particular has exploited stocks of lower-quality, fuel-grade plutonium by blending it with 'super-grade' plutonium containing 3 per cent or less of ^{240}Pu; France has extracted plutonium from uranium blankets inserted in fast reactors; and Britain and France have derived weapon-grade plutonium from 'civil' power reactors.

In this context it is important to note that a small proportion of the plutonium discharged from production reactors is 'lost' during plutonium separation and warhead fabrication, as well as when plutonium is recovered from retired warheads. This occurs because small amounts of plutonium are left behind in chemical residues, or in scraps, some of which may be difficult or uneconomic to recover. The quantities of plutonium lost in this way depend on the efficiency of the production process, and on the trouble weapon manufacturers take in recovering materials from scrap. Plutonium stocks are also diminished by weapon testing, to an extent that depends on the frequency of tests and the sizes and designs of devices being exploded.

III. Methods of estimating military plutonium inventories

There are three standard approaches to estimating the inventories of military plutonium possessed by the NWS: counting warheads, measuring krypton-85 emissions and studying the design, fuelling and operating histories of production reactors.

Counting warheads

This approach involves taking the number of warheads held by a given country, and making assumptions about the average quantity of plutonium per warhead. There are three difficulties here. The first is that only the number of US warheads is known with any precision. As a result of the Strategic Arms Reduction Talks (START), the size of the former USSR's strategic arsenal has become known, but information on its stocks of tactical nuclear weapons remains incomplete. Furthermore, in addition to the Soviet warheads included in the START inventory, several thousand old warheads may be held in store in Russia. In the cases of France and the UK, estimates of warhead numbers are

---

[5] This appears to have reflected a policy towards innovation in nuclear weaponry different from that followed in the USA. In the USSR, warheads were standardized to a greater extent, with the delivery system being adapted to take the warhead. In the USA, many different designs of warheads and associated delivery vehicles were developed.

also subject to large ranges of uncertainty, while in the case of China, estimating warhead numbers is still largely a matter of guesswork.

The second problem is that the amounts of plutonium in nuclear weapons vary widely, depending on design specifications and the quantities of HEU that are used in the fission stages. Even the averages may differ across countries and historical periods. In the following pages the average range of 3–4 kg per warhead is used (except for Chinese and old Soviet warheads, in which the plutonium content is probably lower and the HEU content higher), with the caveat that the actual averages, or 'typical' amounts, could be higher or lower.

The third drawback with this approach is that it only allows estimates to be made of the inventory of plutonium in weapon arsenals, that is, it tells nothing about quantities that are held in store or are locked up in the production process. These quantities can be substantial.

Measuring krypton-85 emissions

Before knowledge was gained of Soviet production facilities, the only way of estimating plutonium output was to measure the buildup of krypton-85 (^{85}Kr) in the atmosphere. By deducting the amounts of ^{85}Kr believed to have come from other countries' nuclear activities, the Soviet contribution could be estimated.

Krypton-85 is a product of the fissioning of ^{235}U or ^{239}Pu in nuclear reactors. It is an inert gas which is retained in reactor fuel until its release during reprocessing. Hitherto, it has been mostly vented into the atmosphere by the reprocessors.[6] The concentration of ^{85}Kr in the atmosphere thus provides an indirect measure of the extent of fissioning in reactor fuel which has been reprocessed, from which estimates of plutonium arisings can be made.

There are two difficulties associated with this method of assessing plutonium inventories. The first concerns the accuracy of measures of ^{85}Kr concentrations in the atmosphere. As ^{85}Kr has a relatively long half-life (10.8 years), global releases can in principle be assessed from local sampling since the gas has time to diffuse widely through the atmosphere. However, measurements have to be taken far away from reprocessing sites, and account has to be taken of varying concentrations and rates of diffusion across the different layers in the atmosphere.[7]

The second difficulty is that the interpretation of ^{85}Kr releases rests on assumptions about the type and operation of production reactors. The rates of fissioning of ^{235}U and ^{239}Pu, and thus of ^{85}Kr production, are functions of reactor fuelling and burn-up. The ^{85}Kr approach thus does not obviate the need for a detailed understanding of production reactors and their operating histories. Traditionally, calculations of plutonium arisings by this method have relied on the very general assumption that US and Soviet production reactors have had

---

[6] Some Western intelligence analysts have speculated that krypton filters have been installed in Soviet reprocessing plants, leading to underestimates of Soviet plutonium output. However, there is no evidence that a significant number of filters were installed.

[7] A full account of the measurement of ^{85}Kr concentrations is contained in von Hippel *et al.* (note 1).

similar fuelling arrangements. They have thus been first approximations. Moreover, account has had to be taken of the significant quantities of ^{85}Kr released by other activities, notably the reprocessing of civil reactor fuels and nuclear weapon testing.

Studying production reactors and operating histories

The best estimates of plutonium inventories come from studying the designs, fuelling arrangements and operating histories of production reactors. While this knowledge is still incomplete, a fuller picture has emerged over the past decade.

In the USA details of the thermal power output (i.e., heat output) of production reactors have been published by the Department of Energy since the early 1980s. Given an understanding of the relationships between thermal output and the fissioning of uranium and plutonium, it became possible to calculate past arisings of plutonium with considerable accuracy. There is one important caveat: in some US reactors, uranium fuel has been replaced on occasion by lithium targets in order to produce tritium for nuclear weapons. Wherever this has occurred, plutonium output has been reduced accordingly. However, although much less precise, information about tritium production has also become available, so that adjustments to calculations of plutonium arisings can be made.

In the case of the UK and France, the main production reactors are dual-purpose electricity and plutonium producers. As their electrical output is on the public record, their thermal output can be calculated given knowledge of the efficiency with which heat is converted to electrical energy. As in the case of the USA, plutonium arisings can then be estimated, although, for reasons explained below, not with the same degree of accuracy.

By comparison, little is known of the operating histories of Soviet and Chinese production reactors. With the exception of at least one Soviet HWR, all are LWGRs, of the type used by the USA at Hanford (see table 3.1). Thus for the USSR, estimates have largely been derived by assuming that its LWGRs exhibited power ratings and burn-ups similar to their counterparts at Hanford. (The extent of Soviet tritium production has also been estimated on the basis of the US experience.) The same approach is adopted for China, but with still less confidence in its reliability. In both cases there is some justification in basing estimates on the US experience, since both the USSR and China learned from, and even copied, US reactor designs.

The estimates of military plutonium inventories for the NWS presented in sections IV–VIII below are derived mainly from the study of production reactors and their operating histories. The one exception is with regard to the USSR, for which a combination of approaches is used. Each section also summarizes what is known about stockpiles of nuclear warheads in the respective country, in order to provide a rough cross-check, and to allow a gross estimate of the amounts of weapon-grade plutonium that are held outside nuclear weapons and would become available when the nuclear weapon states' arsenals

Table 3.2. US production reactors

Reactor site/ designation	Type and power rating (MWth)[a]	Period of operation
Hanford		
B, D, F, H, DR, C	LWGR, up to 2 310	Start-up 1944–52, shut-down 1964–69
KW, KE	LWGR, up to 4 400	Start-up 1955, shut-down 1970–71
N	LWGR, 4 000	Produced weapon-grade Pu 1964–65, 1982–86; fuel-grade Pu 1966–82
Savannah River		
R	HWR, up to 2 800?	Start-up 1953–54; shut-down 1964
L	HWR, up to 2 800?	Start-up 1953–54; shut-down 1968; re-started 1985; shut-down 1988
P, K, C	HWR, up to 2 915[b]	Start-up 1954–55; shut-down 1988

[a] Megawatts-thermal.

[b] The P and K reactors were operated as producers of super-grade plutonium from 1981 (together with the re-started L reactor after 1985) until their closure in 1988. The K reactor may be restarted, at least to demonstrate operational capability. The C reactor was mainly operated as a dedicated tritium producer in the 1980s. Tritium was also produced in varying amounts by these reactors during the 1970s.

Source: Derived from Cochran, T. B, Arkin, W. M., Norris, R. S. and Hoenig, M. M., *Nuclear Weapons Databook, Volume II: US Nuclear Production* (Ballinger: Cambridge, Mass., 1987), pp. 58–65.

of nuclear weapons are reduced through arms reductions. Estimates of plutonium arisings based on the measurement of ^{85}Kr are alluded to only in the context of estimating the total production of plutonium in the former USSR.

IV. The United States

Plutonium production history

The history of weapon-grade plutonium production in the USA falls into three phases.[8] During the first phase, lasting from the mid-1940s to the mid-1960s, up to 14 production reactors were operated: nine at the Hanford Reservation in Washington State, and five at Savannah River in North Carolina (see tables 3.1 and 3.2). Plutonium production reached its peak in the early 1960s, at a rate of around 7 tonnes per annum (figures 3.2 and 3.3). By 1964 over 60 tonnes of weapon-grade plutonium, or approximately two-thirds of the total US inventory, had already accumulated.

By and large, the rate of plutonium production was matched to the anticipated buildup of nuclear warheads. When their numbers peaked in the mid-1960s, and when plutonium began to be recycled from retired warheads, the demand for additional weapon-grade plutonium declined. Between 1964 and

[8] This section draws particularly on Cochran *et al.* (note 1). Readers should refer to this text for a more detailed description of the methodology and data used in calculating plutonium arisings.

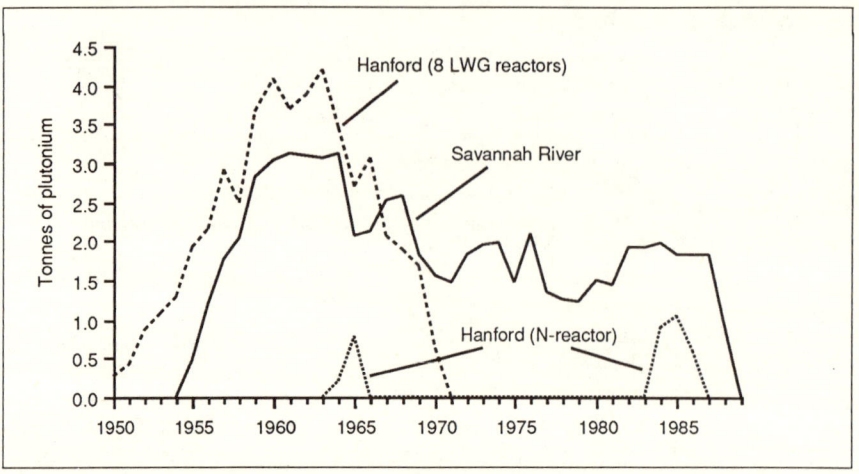

Figure 3.2. US military production by site, 1950–89

Source: Cochran, T. B., Arkin, W. M., Norris, R. S. and Hoenig, M. M., *Nuclear Weapons Databook, Vol. II: US Nuclear Warhead Production* (Ballinger: Cambridge, Mass., 1987), and authors' data.

1971 the original eight Hanford reactors were closed down, and the newly commissioned N-reactor at Hanford was operated to produce electricity at minimum cost. Operating at higher burn-ups, it became a source of fuel-grade plutonium which was expected to be used by an expanding civil fast-reactor programme. The PUREX reprocessing plant at Hanford also ceased operating in 1972. In this second phase, from about 1964 to 1981, the production of weapon-grade plutonium was therefore concentrated at Savannah River; in this period the annual output was 1–2 tonnes of material.

The third phase lasted from 1981 to 1988. It began with ambitious plans to expand plutonium production, as part of the Reagan Administration's arms buildup. In reality, the increases in plutonium output were comparatively modest. The N-reactor at Hanford returned to the production of weapon-grade plutonium in 1982, with the PUREX reprocessing plant reopening in 1984 to handle its fuel. The L-reactor at Savannah River, which had ceased operating in 1968, was also restarted in late 1985. In addition, the P, K and C reactors at Savannah River were adapted to produce super-grade plutonium which was blended with the fuel-grade material that had accumulated in spent fuel from the N-reactor. The stock of fuel-grade plutonium was thereby brought into play, in effect allowing the acquisition of 3 tonnes of weapon-grade plutonium for every 2 tonnes of super-grade plutonium extracted from the Savannah River reactors. Taken together, these steps led to an approximate doubling of the US output of weapon-grade material, the annual production rate approaching 2.5 tonnes during the mid-1980s.

By the end of 1988, however, all US plutonium production had ceased. Safety and environmental problems at the production sites, and the *rapproche-*

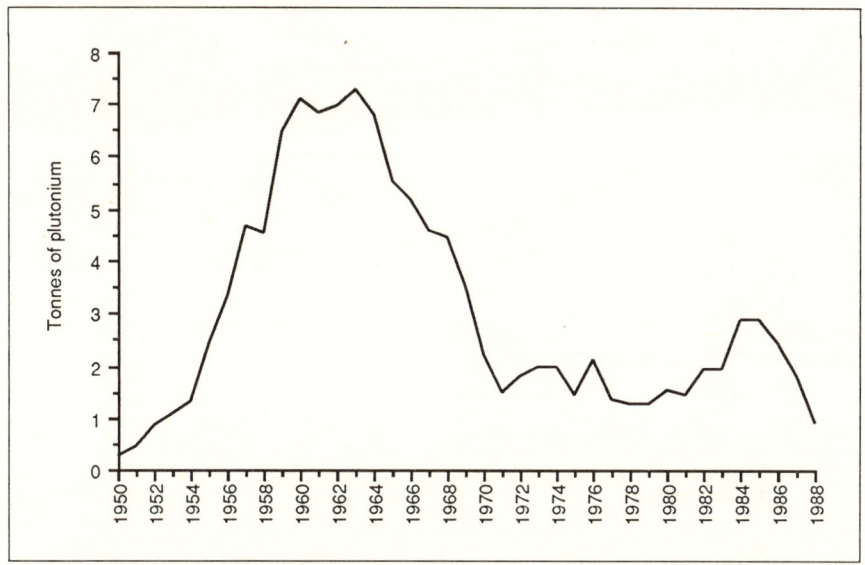

Figure 3.3. Total annual US military production of plutonium, 1950–88

Source: Cochran, T. B., Arkin, W. M., Norris, R. S. and Hoenig, M. M., *Nuclear Weapons Databook, Vol. II: US Nuclear Warhead Production* (Ballinger: Cambridge, Mass., 1987), and authors' data.

ment between East and West, led to the shut-down of the remaining reactors at Hanford and Savannah River. The US Department of Energy announced in early 1991 that it had no plans to produce plutonium 'in the foreseeable future'.

US weapon-grade plutonium

The amounts of weapon-grade plutonium available to the US nuclear weapon programme are shown in table 3.3. These figures are updates of estimates derived by study groups at Princeton University and the Washington-based Natural Resources Defense Council (NRDC) in the mid- to late 1980s, on the basis of information on reactor production histories released by the DOE.[9]

Table 3.3 shows that the plutonium has come from the following sources (it is assumed that virtually all plutonium discharged from production reactors had been separated out by the end of 1990):

1. A total of 49.1 tonnes was produced by the eight Hanford LWGRs, the last of which was shut down in 1971.

2. A total of 3.5 tonnes was produced by the N-reactor at Hanford. Normally optimized to produce electricity, it supplied weapon-grade plutonium in 1964–65 and 1982–86.

3. A total of 48.5 tonnes was produced by the five Savannah River reactors. The thermal power outputs and operating records of these reactors imply a plu-

[9] von Hippel *et al.* and Cochran *et al.* (note 1).

Table 3.3. Estimated US inventory of weapon-grade plutonium, end of 1990
Figures are in tonnes.

Source	Sub-total	Total
Hanford		
8 LWGRs		49.1
N-reactor		3.5
Savannah River		
Plutonium equivalent	66.2	
Tritium (Pu equivalent)	– 16.2	48.5
Other isotopes (Pu equivalent)	– 1.5	
Imports from Britain		1.0
Sub-total		**102.1**
Losses through reprocessing (ca. 2%)		– 2.0
Losses through weapon testing		– 3.0
Total		**97 ± 8%**

tonium output of 66.2 tonnes. However, account has to be taken of the loss of plutonium output due to the production of tritium and other isotopes in these reactors. Cochran *et al.* estimate that 190 kg of tritium had been produced by the end of 1984 (not corrected for decay).[10] An additional 35 kg has been added for the years 1985–88 when one reactor was dedicated to tritium production, giving a total of 225 kg of tritium. This is equivalent to a plutonium output of 16.2 tonnes, assuming that the same amounts of energy are required to produce 1 kg of tritium and 72 kg of plutonium.[11] It is estimated that the equivalent of a further 1.5 tonnes of plutonium production was lost when producing other isotopes in these reactors.

4. The Savannah River total also includes weapon-grade material produced in 1981–88 by blending two parts of super-grade with one part of the fuel-grade plutonium supplied by the N-reactor. It is estimated here that the inventory of weapon-grade plutonium was increased by 2.7 tonnes in this way (which is equal to the amount of fuel-grade plutonium used in blending).

5. An unknown quantity of weapon-grade plutonium was sent from the UK to the USA under the US–UK Mutual Defence Agreement (see section VI below). A nominal amount of 1 tonne has been included here.

This gives a total of 102 tonnes. Taking into account estimated reprocessing losses of 2 tonnes, and the consumption of 3 tonnes of plutonium in weapon tests, this left an estimated US inventory of weapon-grade plutonium of 97 tonnes in 1990, which remained unchanged at the time of writing.

---

[10] Cochran *et al.* (note 1), Appendix C, pp. 179–81.

[11] The term 'plutonium equivalent' is also used in nuclear commerce, where it denotes the amount of ^{239}Pu that has the fissile worth of a given batch of mixed-isotope plutonium.

Table 3.4. Estimated US military inventory of fuel- and reactor-grade plutonium, end of 1990

Source	Tonnes
Fuel-grade plutonium	
Hanford N-reactor	8.1
British Magnox reactors	· 4.0
Other sources	5.0
Material used in blending	– 2.7
Total fuel-grade	**14.4**
Reactor-grade plutonium	0.8
Total fuel- and reactor-grade	**15.2**

It should be stressed that the gross total of 102 tonnes is a central estimate. It is not possible to assess with accuracy the margin of error that should be attached to these calculations. The main potential sources of error lie in the estimates of tritium output, and in the DOE's own measurement of the thermal output of production reactors. Particularly in the 1950s and 1960s, thermal output measures were prone to inaccuracy. The error margin of plus or minus 8 per cent in table 3.3 seems appropriate, and suggests lower and upper bounds of 90 and 106 tonnes respectively.

US fuel- and reactor-grade plutonium

In addition to weapon-grade plutonium, the US Department of Energy possesses around 14 tonnes of fuel-grade and 1 tonne of reactor-grade material (see table 3.4). The former quantity came from two main sources: the N-reactor at Hanford, and British Magnox reactors under the US–UK Mutual Defence Agreement.

During its operation the N-reactor produced a total of 8.1 tonnes of fuel-grade plutonium. In the 1970s it was anticipated that this plutonium would be used in the US fast reactor programme. Since the indefinite deferral of this programme in 1976, most of this plutonium has been held in store. In the 1980s an estimated 2.7 tonnes of this material were blended with super-grade plutonium from the Savannah River reactors.

The UK supplied 4 tonnes of what is believed to have been fuel-grade plutonium to the USA after 1964 (see below). It has been used mainly in the Zero Power Plutonium Reactor (ZPPR) at the Idaho National Engineering Laboratory, and to a lesser extent in the Fast Flux Test Facility (FFTF) at Hanford. The ZPPR plutonium is essentially unirradiated, while the FFTF plutonium is largely contained in spent fuel. These R&D facilities have accounted for 3.8 and 2.9 tonnes of plutonium, respectively.

The remaining 5 tonnes of fuel-grade plutonium arose 'from the accumulation of material from many sources over the past 40 years—for example, from commercial reactor fuel reprocessing operations (West Valley, New York), the accumulation of material from other US government reactor operations, material obtained by barter, and donations from firms and foreign governments.'[12]

Plutonium in US nuclear warheads

At the beginning of 1991 the USA had some 19 000 nuclear warheads deployed in its strategic and tactical arsenals. Numbers had fallen steadily from the peak of around 33 000 in the mid-1960s. Over this period, the average plutonium content rose from 2–3 kg in the 1960s to 3–4 kg in 1991.

Because of the great variety in nuclear warhead size and design, it is not possible to make an accurate assessment of the plutonium inventory contained in US nuclear weapons. The 3–4 kg average, however, leads to an inventory in the range of 57–76 tonnes. This being the case, between 21 and 40 tonnes of weapon-grade material may exist outside weapons. However, a substantial proportion is in the form of unrecovered scrap. According to US Government sources, there are about 10 tonnes of plutonium scrap at the Rocky Flats plant in Colorado, where plutonium pits were manufactured. This means that the surplus inventory of readily usable material may be between 11 and 30 tonnes, giving a central estimate of some 20 tonnes of plutonium. The wide margins of error reflect the paucity of information, even in the USA, about the stocks of weapon material held outside warheads.

In summary, the USA has acquired an estimated 97 tonnes of weapon-grade plutonium, of which perhaps 67 tonnes are now contained in nuclear warheads, 20 tonnes are surplus to requirements, and 10 tonnes are in scraps and other wastes. Of these numbers, only the first carries a margin of error that is less than 10 per cent. The DOE also holds a stock of around 15 tonnes of fuel- and reactor-grade plutonium, of which a little over 2 tonnes has not been separated.

V. The former Soviet Union

Soviet krypton-85 emissions

Until the late 1980s, little was known about Soviet plutonium production facilities. Estimating the Soviet contribution to the buildup of ^{85}Kr in the atmosphere was the only method available for assessing the magnitude of Soviet plutonium production. The most thorough independent study based on this technique was carried out at Princeton University in the mid-1980s.[13] The figures derived for the year 1984 are summarized in table 3.5.

These calculations gave an upper bound of 140 tonnes for the Communist countries. To arrive at a figure for Soviet military plutonium production it is

---

[12] Cochran *et al.* (note 1), p. 76, quoting from a letter from F. C. Gilbert, Acting Deputy Assistant Secretary, Department of Energy, to Thomas B. Cochran, 24 Mar. 1981. For a brief comment on the reactor-grade fuel, see Cochran *et al.* (note 1), p. 75.

[13] von Hippel *et al.* (note 1).

Table 3.5. Estimated Soviet military plutonium output, calculated from estimated krypton releases and plutonium arisings, 1984

Source	Million Curie (mCi)	Tonnes
Krypton (^{85}Kr) releases		
Krypton release in non-Communist countries		
Nuclear weapon tests	5.0	
Reprocessing in USA	47.4	
Reprocessing in UK	18.5	
Reprocessing in France	2.3	
Reprocessing in other West European countries and in Japan	1.7	
Leakage from fuel outside Communist countries	10.7	
Sub-total	**85.6**	
Krypton releases from reprocessing in Communist countries	59.2 ± 10	
Total measured krypton release to atmosphere	$\mathbf{144.8 \pm 10}$	
Plutonium output		
Total plutonium equivalent		140 ± 25
Total plutonium equivalent minus plutonium-equivalent production for other uses and in China		-25
Estimated Soviet military plutonium output		$\mathbf{115 \pm 20}$

Source: von Hippel, F., Albright, D. H. and Levi, B. G., *Quantities of Fissile Materials in US and Soviet Weapon Arsenals*, PU/CEES Report no. 168 (Center for Energy and Environmental Studies, Princeton University: Princeton, N. J., July 1986), table 5.17.

necessary to subtract the outputs of the reprocessing of submarine, research reactor, civil power and tritium fuels in the USSR, and of reprocessing in China. Precise numbers cannot be attached to any of these variables. However, the total output of plutonium in China may have been as little as 2 tonnes, and it is a fair assumption that the amount of reprocessing for alternative uses was proportionately similar in the USSR and the USA.[14] This being the case, the estimated upper bound for the Soviet inventory of military plutonium in 1984 falls to 115 tonnes, with a margin of error of plus or minus 20 tonnes.

Production reactors and the plutonium inventory

Although knowledge about the former USSR's plutonium production system has grown in recent years, much less is known than in the US case. The sites, numbers and types of production reactors are known, but their detailed production histories remain for the most part undisclosed.

---

[14] This is also the assumption made by Cochran *et al.* (note 1). It has since been discovered that the rate of reprocessing water–water power reactor (VVER) fuels was higher than anticipated in the 1980s, giving rise to some 25 t of separated plutonium in 1991. However, the 10 t that could have been separated by 1984 would have been roughly matched by the 8 t of fuel-grade plutonium derived from the N-reactor.

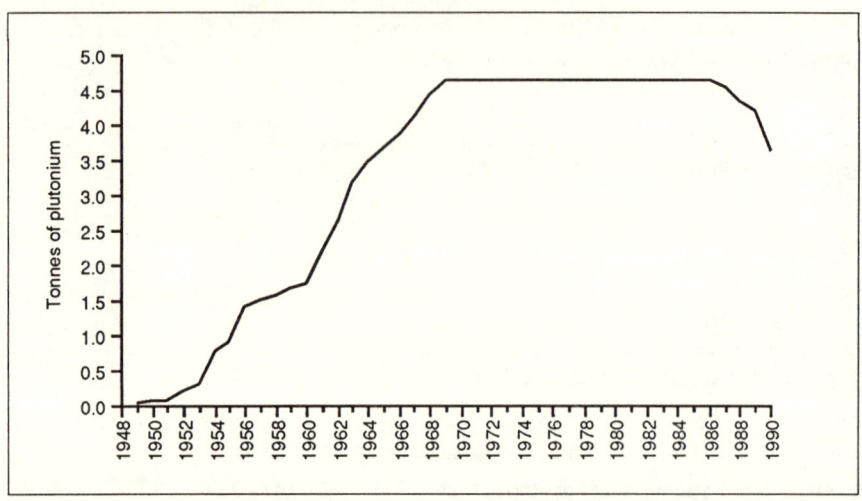

Figure 3.4. Rough profile of Soviet military production of plutonium, 1949–90

Source: Cochran, T. B. and Norris, R. S., *Soviet Nuclear Warhead Production* (Natural Resources Defense Council: Washington, DC, Feb. 1991).

A comparison of figures 3.3 and 3.4 illustrates the difference in historical profiles between US and Soviet plutonium production.[15] Production in the USSR began at around the same time as in the USA, but was built up more slowly. Output peaked towards the end of the 1960s, a decade later than in the USA, and was held at this high level until as late as 1987.

This is consistent with the picture of the USSR's lag in warhead production and deployment in the 1950s and 1960s, and the dogged efforts to match and even surpass the US weapon arsenal in the Brezhnev era. It also suggests a rather inflexible production system which, once in its groove, maintained output at constant levels. Furthermore, it lends support to the view that warheads were not routinely constructed from recycled material.

As table 3.1 shows, there are three production sites in the former Soviet territories (all in Russia): at Chelyabinsk on the eastern side of the Urals, and at Tomsk and Dodonovo (near Krasnoyarsk) in Siberia. The five LWGRs at Chelyabinsk were constructed in the late 1940s and early 1950s, and were all shut down between 1987 and 1990. A single HWR at Chelyabinsk may still operate, if so probably as a tritium producer. In the late 1970s the Mayak reprocessing plant at Chelyabinsk was adapted to process spent fuel from the 440-MW water–water power reactors (VVER) in the Soviet Union, Eastern Europe and Finland. The intention was to use the plutonium arisings in the fast reactors that were also planned for the Chelyabinsk site. It is claimed that by

---

[15] Figure 3.4 is based on the mean of the high and low scenarios of Soviet plutonium equivalent production in Cochran and Norris (note 1), tables 5–7. The graph of Soviet production in figure 3.4 is an approximation: the straight line over the period of maximum output betrays the lack of knowledge about actual operating records.

Table 3.6. Estimated Soviet plutonium equivalent production, ends of 1983 and 1990: low and high scenarios

Figures are in tonnes.

Source	End of 1983	End of 1990
Low scenario		
Chelyabinsk	31	37
Tomsk	38	50
Krasnoyarsk	21	28
Total low scenario	**90**	**115**
High scenario		
Chelyabinsk	43	50
Tomsk	52	70
Krasnoyarsk	30	40
Total high scenario	**125**	**160**

Source: Cochran, T. B. and Norris, R. S., *Soviet Nuclear Warhead Production* (Natural Resources Defense Council: Washington, DC, Feb. 1991), tables 5–7.

1991, when the fast reactor programme had come to a standstill, 25 tonnes of plutonium had been separated from VVER fuel.[16] This number should be regarded with caution as the Mayak plant has had a long history of operational problems.

After 1978 fuel discharged from the Chelyabinsk production reactors was sent to be reprocessed at Tomsk, which also has five LWGRs (two of which were shut down in 1990). These have been operated as dual-purpose electricity and plutonium producers, some of the electricity being consumed by the large uranium enrichment plant that is also located on this site.

Least is known about the third site, at Dodonovo near Krasnoyarsk in Eastern Siberia. There are three LWGRs on the site (not HWRs as originally reported), evidently constructed 250 metres underground. In April 1992 it was announced that they would be shut down. Their sizes and operating histories are not yet known. Spent fuel from 1000-MW power reactors in the USSR (and one in Bulgaria) has been shipped to Krasnoyarsk where construction of a large new reprocessing plant began in the 1980s. Its completion has been much delayed.

Of these 14 production reactors, the performance characteristics of only three at Chelyabinsk are known in any detail. The figures for plutonium output provided by Cochran and Norris are therefore much more conjectural than for the USA. The extent to which reactors have been used to produce tritium is also unknown. Cochran's and Norris's estimates of plutonium equivalent production from the three sites is summarized in table 3.6.

For the USA, the amount of plutonium available to the weapon programme was approximately 20 per cent less than the amount contained in fuels from production reactors, after allowance was made for tritium production, reproces-

[16] Bukharin, O., 'Soviet reprocessing and waste management strategies', Paper presented at Princeton University's Workshop on Nuclear Weapons Materials Disposal, Berlin, 28–30 Nov. 1991.

sing losses, weapon testing and so on. Deducting the same proportion for the USSR gives ranges of 72–100 and 92–128 tonnes at the ends of 1983 and 1990 respectively. The figures for 1983 are considerably below the central estimate of 115 tonnes arrived at by estimating Soviet ^{85}Kr emissions. Using Cochran's and Norris's assumptions about the rate of production later in the 1980s, this method would produce a figure of around 140 tonnes for the latter period.

At the beginning of 1992 six of these production reactors were believed still to be operating and their discharged fuel being reprocessed. The production system had not yet closed down in Russia, as it had in the USA. Russian officials claimed that the reactors were being maintained to supply heat and electricity, although there have been indications that weapon-grade plutonium is still being produced in the Tomsk reactors, which alone operate today.

In 1989 the Soviet official in charge of nuclear materials production in the Ministry of Atomic Power and Industry (MAPI) said that the USSR had produced 'a little bit more' plutonium than the USA.[17] The question is how much more. How tongue-in-cheek was this remark? At present there is no way of telling, but the large number of warheads in the Soviet nuclear arsenal suggests that the amount produced may be higher than has sometimes been assumed.

Plutonium in the CIS nuclear warheads

Estimates of the number of nuclear warheads formerly deployed by the former USSR have varied greatly. The number of warheads on strategic weapons was declared to be 10 271 when the START Treaty was signed in July 1991. In 1992 the total number of warheads is considered to lie towards the top end of the range 27 000–34 000.[18] If there are 17 000–18 000 tactical weapons, as has recently been suggested, as many as 7000 old warheads may be held in store.

Assuming that there are 27 000 operational warheads and 5000 in store, and that they contain, on average, 3–4 kg and 2–3 kg of plutonium per warhead, respectively, the total complement of warheads would contain 91–123 tonnes of plutonium. However, the quantity of material required to support a buildup to this level of armament will have been significantly higher, especially if weapon production had been marked by the kind of inefficiency found elsewhere in the Soviet industrial system. Assuming that 10 per cent of plutonium arisings are contained in unrecovered scraps and are lost during reprocessing, and that a further 10 per cent above overall warhead requirements is needed to maintain production rates, the total inventory required increases to the range 109–148 tonnes, giving a mean requirement of 128 tonnes. It is not known whether a significant proportion of the plutonium coming from production sites has been fuel or reactor grade, and thus inappropriate for warhead manufacture.

[17] Cochran and Norris (note 1), p. 32.

[18] Viktor Mikhailov, head of the Russian nuclear weapon industry, has said that 27 000 warheads is the 'lowest estimate', but that it is accurate to 'within 15–20 per cent'. This suggests an upper limit of 32 500 warheads. Other reports have placed the number at 34 000. See 'Russia says it has too many warheads to cope', *International Herald Tribune*, 5 Feb. 1992.

Evaluating the various estimates presented above would require information that has not yet been made available. A central estimate of 125 plus or minus 25 tonnes for the size of the Soviet military plutonium inventory at the end of 1990 seems the best that is possible today. The large error margin is unavoidable given the remaining uncertainty surrounding nuclear weapon production in the former USSR.

VI. The United Kingdom

While the British, French and Chinese military plutonium inventories are much smaller than the US and Soviet inventories, reflecting the relatively modest scale of their nuclear weapon programmes, they are no easier to assess.

The main plutonium production site in the UK is at Windscale, subsequently re-named Sellafield, in the north-west of England. The Calder Hall reactors and the plutonium separation plants are located there. The Chapelcross reactors are situated near Dumfries, in the south-west of Scotland. Plutonium was derived from the two Windscale piles between 1951 and 1957, and after 1957 from the Calder Hall and Chapelcross production reactors.

Simpson estimates that the Windscale piles produced around 400 kg of plutonium in total. Calder Hall and Chapelcross were operated as suppliers of weapon-grade plutonium between 1957 and 1964, after which they were optimized to produce electricity, thereafter producing mainly fuel-grade plutonium. On one occasion after 1964—perhaps in 1979—weapon-grade plutonium was again supplied from these reactors. Two of the eight reactor units at Calder Hall and Chapelcross were refuelled and operated at burn-ups of 800 MWd/t, providing an estimated additional 170 kg of weapon-grade material (or 0.2 tonne when rounded up).[19]

The output of weapon-grade plutonium from Calder Hall and Chapelcross is an estimated 3.4 tonnes, the output in the period 1957–64 being 3.2 tonnes. This last figure is calculated from an electrical output of 19 terawatt-hours (TWh), and plutonium arisings of 0.36 kg per tonne at an average burn-up of 400 MWd/t.[20]

The other sources of weapon-grade plutonium were the first discharges of fuel from the 'civil' electricity-producing reactors constructed in the 1960s. A little history is required to explain the origin of this plutonium, and of the material sent to the USA. Until 1969 reprocessing in the UK was conducted by the United Kingdom Atomic Energy Authority (UKAEA), which retained

[19] Simpson (note 1), p. 252, table A4a.

[20] To the end of 1966 electricity production from these reactors was 25.4 TWh; see *Nucleonics Week*, 19 Jan. 1967. Simpson (note 1) estimates that production in 1965–66 amounted to 6.3 TWeh. 19 TWh is equivalent to 3520 gigawatt-days of thermal energy (GWdth), the efficiency of converting thermal to electrical energy in these reactors being 22.5%. 3520 GWdth yields 8800 t of irradiated fuel at 400 MWd/t, containing 3.17 t of plutonium at 0.36 kg/t (conversion rates are from Turner, S. E., *Criticality Studies of Graphite-Moderated Production Reactors*, Report prepared for the US Arms Control and Disarmament Agency, document no. SSA-125 (Southern Sciences Applications Inc., Jan. 1980)). Taking a different and in our view less reliable route (relating plutonium arisings to annual thermal output achieved at a given load-factor), Simpson arrives at the lower estimate of 2.34 t over this period.

ownership of all British plutonium, whether derived from civil or military reactor fuels. In 1969 and in 1971 the two utilities operating nuclear power stations—the Central Electricity Generating Board, covering England and Wales, and the South of Scotland Electricity Board—gained title to the plutonium henceforth separated from their civil spent fuels.

Before 1969 no clear distinction was made between civil and military plutonium in the UK. The main distinction was drawn between weapon- and reactor-grade plutonium (the latter including fuel-grade in the British terminology).[21] All weapon-grade material was made available by the UKAEA to the Ministry of Defence, including, it is presumed, the plutonium derived from the initial low burn-up discharges from the seven civil Magnox reactors commissioned before 1969. These discharges would have yielded around 300 kg of weapon-grade plutonium.[22] Until the mid-1980s, when it became the policy to run separate campaigns, spent fuels from civil and military reactors were reprocessed together ('co-processed').

Under the US–UK Mutual Defence Agreement of 1958, as amended in 1959, British plutonium was bartered for US highly enriched uranium and tritium.[23] The precise terms of the exchange have not been revealed, but there were two arrangements covering plutonium, one for weapon-grade and the other for fuel-grade material. The export of plutonium to the USA has been controversial in the UK, the claim being made by some critics that material from British civil reactors has been used in the US weapon programme.[24] What material was dispatched when and from which source remains unclear, but what is known may be summarized as follows:

1. Between 1959 and 1964 weapon-grade plutonium from the Calder Hall and Chapelcross production reactors was exported to the USA.

2. Between 1964 and 1971 approximately 4 tonnes of fuel-grade plutonium were exported, most and possibly all of which derived from civil reactor fuels.

3. Between 1971 and 1984 only 50 kg of fuel-grade plutonium was transferred to the USA, of which 40 kg came from Chapelcross and Calder Hall. The main transfers, particularly between 1976 and 1979, seem to have involved weapon-grade plutonium from pre-1971 UKAEA military stock.

The total amount of weapon-grade plutonium exported to the USA is still not known. Simpson suggests that 600 kg may have been transferred prior to

---

[21] Most plutonium extracted from Magnox reactor fuels belongs in the category 'fuel-grade'.

[22] This assumes that the first one-sixth of a core, or 665 t of fuel in seven reactors, is submitted to a burn-up of 500 MWd/t, which gives rise to 0.45 kg of plutonium per tonne.

[23] A detailed examination of this agreement has been provided by Simpson (note 1). During the 1960s and 1970s the availability of sufficient and reliable reprocessing capacity to provide plutonium for the British weapon programme and for export to the USA, and to reprocess fuels from the growing stock of civil reactors, was the main problem faced by British authorities.

[24] See the exchanges on 7, 8, 9 and 28 Nov. 1984 at the Sizewell B Public Inquiry. These were partly initiated by an analysis of plutonium arisings which suggested that some 2 t of plutonium were 'unaccounted for'. See Barnham, K. W. J., Hart, D., Nelson, J. and Stevens, R. A., 'Production and destination of British civil plutonium', *Nature*, vol. 317, no. 6034 (19 Sep. 1985), pp. 213–17.

Table 3.7. Estimated British inventory of military plutonium, end of 1990.

Source	Tonnes
Windscale piles	0.4
Calder Hall and Chapelcross	3.4
Initial discharges from civil reactors	0.3
Subtotal	**4.1**
Losses through reprocessing (ca. 2%)	− 0.1
Losses through weapon testing	− 0.2
Exports to the USA	− 1.0
Total	**2.8 ± 0.7**
Fuel-grade plutonium in military inventory	7.6

1964.[25] As indicated in table 3.3, a nominal figure of 1 tonne, which the authors have been given reason to believe may be close to the mark, has been attached to this export. The actual amount could nevertheless be considerably higher or lower. It is also possible that some of the material allocated to the USA under the Mutual Defence Agreement has remained in the UK, or that some of this material or of the material transferred to the USA has subsequently been loaned or reallocated to the UK (particularly to assist the Trident warhead programme). Some of the weapon-grade material exported before 1964 came to form part of the 4 tonnes of stock used in the FFTF and ZPPR in the USA.

The estimated British inventory of weapon-grade plutonium is summarized in table 3.7. According to the central estimate, 2.8 plus or minus 0.7 tonnes have been available to the British weapon programme, the main uncertainty being the scale of exports to (or subsequent re-imports from) the USA.[26] In addition, an estimated 8.3 tonnes of fuel-grade plutonium had been produced at Calder Hall and Chapelcross from 1964 up to the end of 1990, of which 0.7 tonne is known to have been exported to countries other than the USA.[27] This fuel-grade plutonium is still categorized as military material, and remains outside international safeguards.

A total of 2.8 tonnes of weapon-grade plutonium does not leave the UK with a substantial margin above its weapon needs. The size of the British nuclear arsenal is a closely guarded secret, but is believed to lie in the range 200–300 warheads. The number may rise to around 500 when the Trident submarine

[25] Simpson (note 1), p. 262.

[26] As the amount of tritium used or produced for weapons in the UK is unknown, no attempt has been made to take tritium into account when deriving the plutonium estimate. If the very rough assumption were made that enough tritium was needed to boost 300 warheads over a 20-year period, an initial inventory of 2 kg and an annual addition of 100 g might suffice, but such amounts would not likely have affected the estimate presented here of plutonium produced at Calder Hall and Chapelcross up to the end of 1964. Whether tritium production has substituted for plutonium production depends on the method used. Since 1980 any plutonium production foregone will have been fuel-grade, as the main tritium production has since occurred at Chapelcross, which has been optimized to produce electricity. Some tritium is also believed to have been imported from the USA.

[27] Evidence given by Mr J. Baker on 9 Nov. 1984 to the Sizewell B Public Inquiry. The precise amount is 663 kg.

fleet comes into operation in the mid-1990s.[28] At 3–4 kg per warhead, this would require 1.5 –2 tonnes of plutonium.

VII. France

France has derived weapon-grade plutonium from a multiplicity of sources: the three dual-purpose 'G' reactors (GCRs) and the twin Celestin HWRs at Marcoule; the Magnox power reactors, particularly at Chinon; the Rapsodie and Phénix prototype breeder reactors at Marcoule; and possibly Spain's Vandellos Magnox reactor which was built by France in the 1960s, the spent fuels from which have been returned to France. The irradiated fuels from these sources have been reprocessed in the UP1 plant at Marcoule and the UP2 plant at La Hague. At no time has France made a distinction between civil and military plutonium. The civil and military programmes are overseen by the Commissariat à l'Énergie Atomique (CEA), whose subsidiary, the Compagnie Générale des Matières Nucléaires (Cogema) provides all relevant fuel-cycle services.

The French nuclear weapon programme formally began in 1958. The estimated amounts of weapon-grade plutonium that have been produced in France are set out in table 3.8. The figures add up as follows:

1. From the 38-MWth G1 reactor, operated at a burn-up of 100 MWd/t between 1956 and 1968: 100 kg.[29]

2. From the 260 MWth G2 and G3 reactors, operating from the late 1950s to 1980 and to 1984, respectively: 2.4–2.7 tonnes.[30] Their burn-ups have varied from 100 to 1200 MWd/t, depending on the quality of plutonium being sought, and the degree to which they have been optimized to supply electricity.[31]

3. From the Celestin reactors: 0.5–1.5 tonnes.[32] These reactors have been used mainly to provide tritium but are believed to have produced plutonium from 1980 onwards. At their original power rating of 190 MWth, they are capable of supplying 100 kg of plutonium per annum. However, there is speculation that they were upgraded during the 1980s.[33]

4. From France's prototype fast breeder reactors (principally Phénix): 0.9–1.6 tonnes. Up to the end of 1990 Phénix had provided 21.6 TWh of electricity, which potentially yields 860 kg of super-grade plutonium in the ^{238}U blanket.[34]

[28] The British Polaris missiles are believed to carry two warheads each. With three submarines, each armed with 16 missiles, this gives rise to 96 warheads. The remaining British weapons are free-fall bombs and nuclear depth charges. The maximum feasible Trident deployment is four boats with 16 missiles and eight warheads each, giving 512 warheads. However, it is likely that fewer warheads will be deployed in practice.
[29] Albright, D., *World Inventories of Plutonium*, PU/CEES Report no. 195 (Center for Energy and Environmental Studies, Princeton University: Princeton, N. J., June 1987).
[30] Albright (note 29).
[31] The average burn-ups and estimated spent fuel outputs of these reactors are shown in 'Le dossier electronucléaire', *Syndicat CFDT de l'énergie atomique* (Editions du Seuil: Paris, 1980).
[32] Albright (note 29).
[33] Gsponer, A., Independent Scientific Research Institute, *The French Military Nuclear Fuel-Cycle*, document no. ISRI 82-03 (ISRI: Geneva, Dec. 1983), 3rd version.
[34] Assuming a thermal efficiency of 40%, 0.365 kg of plutonium per MWe-year equates to 41.7 kg per TWh, which gives 0.4 g of plutonium per MWth-days. See Albright (note 29).

Table 3.8. Estimated French inventory of military plutonium, end of 1990

Source	Tonnes
G1, G2 and G3 Magnox reactors	2.5 – 2.8
Celestin 1 and 2 (HWRs)	0.5 – 1.5
Rapsodie and Phénix (FBRs), after blending	0.9 – 1.6
Chinon 1, 2 and 3, and initial loadings in other Magnox reactors	1.1 – 2.3
Sub-total	**5.0 – 8.2**
Losses through reprocessing (ca. 2%)	– 0.1
Losses through weapon testing	– 0.5
Total inventory of weapon–grade plutonium	4.4 – 7.6
Total	**6.0 ± 1.5**

This material may have been blended with fuel-grade plutonium to increase the amount of weapon-grade material. It is not known whether plutonium from the Superphénix reactor has been extracted and made available to the military programme.

5. From the Chinon 1, 2 and 3 reactors, and from initial loadings in the Saint Laurent 1 and 2, Bugey 1 and Vandellos reactors: 1.1–2.3 tonnes. Pharabod suggested that up to 3 tonnes of weapon-grade plutonium has come from these reactors.[35] This seems excessive, particularly since he underestimates the degree to which reactors other than Chinon 1 provided fuel-grade plutonium. It is assumed here that Chinon 1 produced 0.4 tonne of weapon-grade plutonium before it was shut down in 1973, that Chinon 2 and 3 produced 0.5–1.7 tonnes of weapon-grade material, and that at most 0.2 tonne of plutonium was extracted from the initial loadings (one-sixth core) in the remaining Magnox reactors.

6. The relatively high losses through weapon tests reflect the relatively large number of tests (183 up to the end of 1991) that France has conducted. As in the case of the UK, no allowance has been made for possible plutonium losses resulting from tritium production, since the extent and methods of tritium production in France are not known.

In sum, France's inventory of weapon-grade plutonium at the end of 1990 comprised an estimated 6.0 plus or minus 1.5 tonnes. If this is the case, France has produced substantially more material than has been required by its weapon programme. Recent assessments suggest that France has deployed just over 500 warheads.[36] Assuming an average of 3–4 kg per warhead, this implies a requirement for 1.5–2.0 tonnes of plutonium. France may therefore possess a surplus stock of 4–5 tonnes of weapon-grade plutonium.

---

[35] Pharabod (note 1).
[36] See Norris, R. S., Fieldhouse, R. W., Cochran, T. B. and Arkin, W. M., 'Nuclear weapons', SIPRI, *SIPRI Yearbook 1991: World Armaments and Disarmament* (Oxford University Press: Oxford, 1991), table 1.6, p. 23.

VIII. China

Remarkable amounts of information have become available about the Chinese nuclear weapon programme in the past few years, as a result of Chinese publications and the work of Lewis and Xue.[37] Knowledge of plutonium production nevertheless remains hazy. Before Soviet technical assistance was withdrawn in 1960, some design drawings and a few items of equipment for a production reactor were transferred to China. Priority was given to the enrichment programme, and the first production reactor began operating in 1966, two years after the first Chinese weapon test. This reactor, at the Jiuquan Atomic Energy Complex in Subei county (initially code-named Plant 404), experienced frequent technical difficulties, and interruptions because of the political turmoil in China during the Cultural Revolution, and was shut down for repair between 1973 and 1975. In the mid-1980s it was re-engineered so that it could be operated as a dual-purpose electricity and plutonium producer. All fuels from this reactor were reprocessed on site from 1968 onwards.

The other production site (Plant 821) is located at Guangyuan, in Sichuan Province. Its development began in the mid-1960s as a 'third-line' weapon manufacturing facility, further away than Jiuquan from China's frontiers. It is known that China's largest production reactor was constructed on this site. Although there are reports of other reactors in China, there is no evidence that there are plutonium production reactors other than the single units at Jiuquan and Guangyuan.[38]

Lewis and Xue report that the reprocessing plant at Jiuquan was capable of processing 400 kg of uranium fuel daily, and operated for 250 days in each year. This implies a production reactor with a capacity of around 400 MWth and an annual discharge of 100 tonnes of uranium, or approximately 100 kg of plutonium (roughly equivalent to the Soviet A-reactor). Assuming that this reactor was optimized to produce weapon-grade material and operated at half-load on average between 1967 and 1972, and at full-load between 1975 and 1983, the plutonium output would have been a little over 1 tonne.

The production reactor located at Guangyuan is larger. How much larger is not known. Assuming it produced 150–200 kg per year over a 10–15 year period, it may have given rise to 1.5–3 tonnes of plutonium.

According to a Chinese source, even after 1975 the operating performance of the Chinese production reactors was often unsatisfactory. A lower limit of 1 tonne and an upper limit of 4 tonnes is therefore attached to the Chinese inventory of weapon-grade plutonium, giving a central 'guestimate' of 2.5 plus or minus 1.5 tonnes. Until hard information about the production reactors becomes available, these must be regarded as very tentative figures.

[37] The main sources on China are Lewis, J. W and Xue, L., *China Builds the Bomb* (Stanford University Press: Stanford, Calif., 1988); Lewis, J. W. and Xue, L., 'Chinese strategic weapons and the plutonium option', *Critical Technologies Newsletter* (US Department of Energy: Washington, DC, Apr.–May 1988); Fieldhouse R. W., 'Chinese nuclear weapons; a current and historical overview', NWD 91-1 (Natural Resources Defense Council: Washington, DC, Mar. 1991).

[38] Lewis and Xue, *China Builds the Bomb* (note 37), p. 113.

4. Weapon-grade uranium inventories in nuclear weapon states

I. Introduction

Highly enriched uranium in nuclear weapons usually has ^{235}U concentrations in excess of 90 per cent. This material is often referred to as 'weapon-grade uranium'.

The military inventories of weapon-grade uranium in the five acknowledged NWS are largely unknown. Despite years of attempts to understand them, only the US inventory has been estimated in detail.[1] A major reason is that the five NWS built enrichment facilities that are much larger than needed to make weapon-grade uranium for weapons. As a result, estimates of total inventories require knowledge of both historical enrichment production and of non-weapon uses of enriched uranium, such as use in submarine fuel. Only the USA has released historical enrichment production information at its military plants, although it has not provided information about the fraction dedicated to non-weapon uses.

Despite these shortcomings, this chapter provides an overview of the enrichment programmes in the NWS, and presents first-order estimates of the weapon-grade uranium inventories in each of them.

II. Overview of enrichment programmes

The birth of the uranium enrichment industry occurred in the US Manhattan Project during World War II, when the USA developed several approaches to produce the HEU which destroyed Hiroshima. The only techniques that proved capable of producing weapon-grade uranium reliably were the electromagnetic isotope separation (EMIS) and the gaseous diffusion processes. Because large-scale production using EMIS required enormous amounts of electricity, the USA concentrated on building gaseous diffusion plants after the war.

The Soviet Union launched its own enrichment programme during the later stages of World War II. It first deployed gaseous diffusion at the end of the 1940s, and later it built gas centrifuge plants. China received assistance from

---

[1] von Hippel, F., Albright, D. and Levi, B., *Quantities of Fissile Materials in US and Soviet Nuclear Weapons Arsenals*, PU/CEES Report no. 168 (Center for Energy and Environmental Studies, Princeton University: Princeton, N. J., July 1986); and Cochran, T., Arkin, W., Norris, R. and Hoenig, M. M., *Nuclear Weapons Databook, Vol. II: US Nuclear Warhead Production* (Ballinger: Cambridge, Mass., 1987).

Table 4.1. Military enrichment plants in the nuclear weapon states

| Country | Maximum capacity for weapons[a] | | Main production period[d] | Halt to WGU production for weapons | Weapon inventory and reserves (t) |
	SWU/y.[b] (x 10^6)	WGU (t/y.)[c]			
USA	16.5	85	1950s–60s	1964	500–600
USSR	13	65	1970s–80s	1987	520–920
UK	0.4–0.6	2–3	1950s–60s	1963[e]	5–15[f]
France	0.3–0.6	1.5–3.0	1970s–80s	..	10–20
China	0.18–0.3	1.0–1.5	1970s–80s	1987	10–20

[a] Except for USA, these capacity figures are estimates. They represent capacities that could have been dedicated to producing weapon-grade uranium (WGU). Actual production would have been less, because of requirements for naval, production and civil reactors.
[b] Kilogram separative work units per year.
[c] Weapon-grade uranium (t/y.).
[d] Estimated period when production of most WGU occurred.
[e] Refers to domestic production only, does not reflect imports of WGU from the USA.
[f] Includes WGU imported from the USA in exchange for plutonium (see chapter 3).

the USSR in building its gaseous diffusion plant, which began operation in the 1960s. When the UK and France started their nuclear weapon programmes, they also built gaseous diffusion plants. The British plant started in the mid-1950s and the French plant in the late 1960s. The UK also received a significant amount of weapon-grade uranium from the USA.

Nuclear weapons provided the primary motivation for these enrichment programmes, summarized in table 4.1. In all the NWS these programmes were complemented by plutonium production programmes. The NWS have used both fissile materials, either alone or in combination, in their nuclear weapons; but weapon designers clearly desired weapon-grade uranium for the second stages of thermonuclear weapons, where it is used in conjunction with the thermonuclear materials.

III. The United States

The USA curtailed production of weapon-grade uranium for nuclear weapons in 1964 after 20 years of production. Although the US Government planned to resume production in the early 1990s, claiming that its stockpile was running low, it cancelled these plans in early 1991. With the end of the cold war the US nuclear weapon arsenal is undergoing deep reductions, creating large excess stocks of weapon-grade uranium from dismantled weapons.

In November 1991 the USA announced that it will suspend production of HEU for any purpose.[2] Future requirements for naval, production and research reactors can be met from existing stockpiles.

---

[2] Dizard, W., 'Suspension of HEU production viewed favourably by friends, foes of UEE Bill', *Nuclear Fuel*, 25 Nov. 1991.

The US enrichment complex

In the course of the Manhattan Project, the USA developed several enrichment processes, principally electromagnetic separation with calutrons and gaseous diffusion. After the war only gaseous diffusion plants were built, because of their lower cost and greater energy efficiency than those of the other methods available at that time.

The first gaseous diffusion plants were built in Oak Ridge, Tennessee. By the mid-1950s, plants were also located in Paducah, Kentucky and Portsmouth, Ohio. US production rose sharply in the 1950s, and reached a peak of 16.5 million SWU in 1961, out of a total US capacity of 17.2 million SWU per year. At peak production, the USA produced over 80 000 kg of weapon-grade uranium per year.

After production for weapons was terminated, production dropped to a low of about 6 million SWU in 1970. The enrichment plants continued to produce weapon-grade uranium for naval reactors (see below), but they produced mostly low-enriched uranium for civil purposes.

After 1970 production increased again in anticipation of large needs for civilian nuclear power reactors. The enrichment plants were expanded to a total capacity of 27 million SWU per year, but new reactor construction and enrichment sales were significantly lower than expected, and the Oak Ridge plant was shut down in 1985. Construction of the Gas Centrifuge Enrichment Plant at Portsmouth was also cancelled in 1985, after almost $3 billion had been spent.

Weapon-grade uranium production for weapons

The amount of weapon-grade uranium produced for weapons has been estimated by two research groups in the USA, which arrived at slightly different, but overlapping estimates.[3] The NRDC group estimated a weapon stockpile of about 500–530 tonnes; and the Princeton group estimated 550 tonnes devoted to nuclear weapons, and posited another stockpile of 50 tonnes assigned to non-military uses.

Both of these estimates relied almost exclusively on the public record. The NRDC group, however, was told by a Government source that the inventory dedicated to weapons was about 500 tonnes. Their major findings are summarized below .

Historical production

From World War II until 1964 almost the entire US enrichment capacity was dedicated to the production of weapon-grade uranium, and most of this was for weapons. Civil nuclear power was still in its infancy, and thus civil demand was small.

[3] See von Hippel *et al.* (note 1) and Cochran *et al.* (note 1).

Until production ceased for weapons, the enrichment complex produced about 150 million SWU, at an average tails assay of about 0.3 per cent. About 4.6 million SWU of this total went towards producing low-enriched uranium for plutonium production reactors and early power reactors. The rest went to the production of weapon-grade uranium, resulting in the production of an equivalent of about 720 tonnes of weapon-grade uranium.

Of this amount, roughly 125–140 tonnes (25–28 million SWU) went to various non-weapon purposes. Almost half of the draw-downs in the stockpile listed in table 4.1 occurred after 1964. Most of these were to provide fuel for research reactors and the Savannah River production reactors.

Weapon-grade uranium inventories

Subtracting the above draw-downs leaves about 600 tonnes of weapon-grade uranium available for weapons and other purposes. The NRDC group suggests that this estimate was an upper bound on the inventory left for weapons in 1964. They considered that there had been additional draw-downs and inventories, and that a better estimate of the inventory dedicated to weapons was about 500–530 tonnes.

The Princeton group assumed the existence of an unidentified stockpile of non-weapons weapon-grade uranium of about 50 tonnes. This amount of material could also be held in a strategic reserve. This left about 550 tonnes for weapons.

The uncertainty in the Princeton and NRDC estimates was put at about 10 per cent. That uncertainty could be higher.

Not all of this weapon-grade uranium could be in the weapons themselves. Building and dismantling weapons is a large-scale industrial process whose successful operation requires nuclear material in the production 'pipeline' and in strategic reserves. Little information exists about the amount tied up in the production processes, although it could be as high as 20 per cent of the total, or about 100 tonnes.

If 500 tonnes of weapon-grade uranium were contained in weapons by 1965, when the arsenal had an estimated 32 000 nuclear warheads, each warhead would have contained an average of about 16 kg of weapon-grade uranium. This material, however, was contained in both the secondary and the primary stages of weapons. In the latter, the weapon-grade uranium was typically used in conjunction with plutonium. Although plutonium is considered the better material to use in primaries, plutonium production has historically lagged behind uranium production. In 1965, the amount of plutonium per warhead averaged only 2 kg; but currently it averages about 3–4 kg per weapon (see chapter 3).

Since the size of the arsenal has shrunk since the mid-1960s, and the design of the primaries has shifted to plutonium only, the average amount of weapon-

Table 4.2. Independent estimates of US consumption of weapon-grade uranium

Figures are in tonnes.

	Princeton[a] estimates	NRDC[b] estimates
Fuel for the Savannah River reactors (until shutdown in the late 1980s)	40	44
Fuel for naval propulsion reactors (until 1964)	11	13
Fuel for domestic and foreign research reactors (until 1990)	25	25
Export to UK and France for military use	6	5
Consumption in US nuclear weapon tests	20	20
Process losses (3% of original enrichment output)	23[c]	..
In process at enrichment plants	..	7[c]
Working inventory	..	25[c]
Total	**125**	**139**

[a] Center for Energy and Environmental Studies, Princeton University, Princeton, N. J.

[b] Natural Resources Defense Council, Washington, DC. (NRDC give some values as ranges, but only mid-points are listed here.)

[c] These numbers represent the amount of weapon-grade uranium that could have been produced with the enrichment work assigned to this category.

Sources: von Hippel, F., Albright, D. H. and Lei, B. G., *Quantities of Fissile Materials in US and Soviet Weapon Arsenals*, PU/CEES Report no. 168 (CEES, Princeton University: Princeton, N. J., July 1986); Cochran, T. B, Arkin, W. M., Norris, R. S. and Hoenig, M. M., *Nuclear Weapons Databook, Volume II: US Nuclear Production* (Ballinger: Cambridge, Mass., 1987).

grade uranium in a weapon is thought to have decreased.[4] At the end of 1991, the size of the US arsenal was estimated by NRDC to be about 19 000 weapons. If the average amount of weapon-grade uranium in each weapon is about 15 kg, about 285 tonnes are contained in those weapons. This implies that the USA already has a large surplus of weapon-grade uranium.

Weapon-grade uranium produced for naval reactors

After 1964, the USA continued to produce large quantities of weapon-grade uranium (in this case 97 per cent enriched) for naval reactors. Since the *Nautilus* first went to sea in the mid-1950s, US naval nuclear-powered ships have steamed over 135 million km and have operated over 3800 reactor-years.[5] In 1991, the USA had 173 naval reactors in about 145 submarines and surface ships, and in land-based prototypes.[6]

---

[4] Some new weapons, such as the high-yield W88 warhead for the Trident II missile, might use more weapon-grade uranium in the secondary than their predecessors. In this case, weapon-grade uranium might replace lithium deuteride, which is less dense, allowing a reduction in the size of the warhead.

[5] Testimony of Admiral Bruce DeMars, US Navy Director, Naval Nuclear Propulsion, before the Subcommittee on Energy and Water Development, in *Energy and Water Development Appropriations for 1992*, Hearings before the Subcommittee on Energy and Water Development, Committee on Appropriations, US House of Representatives, 102nd Congress (US Government Printing Office: Washington, DC, 1991), Part 6, p. 871.

[6] See testimony of Admiral DeMars (note 5), p. 839.

To fuel its reactors, the Navy procured about 600 fresh nuclear reactor cores up to the end of fiscal year 1984.[7] Because each core requires about five to seven years to produce, cores procured in 1984 would not be loaded on to a ship until about 1990. Because cores take so long to fabricate, they are typically ordered before the ship. In the present period of sharp reductions in military forces, a surplus of cores could develop.

NRDC has estimated that the first 200 cores contained about 80 kg of weapon-grade uranium, and modern ones have an average of 200 kg of weapon-grade uranium.[8] The increase in fuel loadings has resulted from developing cores able to last longer. Early ballistic missile submarines had cores that lasted five years; modern ones can last over 20 years. Applying NRDC's estimates of core size, 600 cores would have required more than 48 tonnes and less than 96 tonnes of weapon-grade uranium.[9] We take the mid-point of this range—72 tonnes—as the total amount of weapon-grade uranium loaded into cores for naval reactors until the end of 1990.

During the mid- to late 1980s, the US Navy required about 3 to 5 tonnes of weapon-grade uranium a year, giving a total requirement of about 18 to 30 tonnes from 1984 until the end of 1990.[10] Most of this material would have been used to make the cores ordered several years earlier, and it is not added to our estimate.

Current allocation of weapon-grade uranium.

If each of the 173 operating reactors had 200 kg of weapon-grade uranium initially, roughly 35 tonnes were loaded into those reactors.

As of early 1991, 267 refuellings of reactors had occurred.[11] These are older cores, which we assume contained initially an average of about 100 kg of weapon-grade uranium. If about 50 per cent of the ^{235}U was consumed, the spent fuel contained about 13 tonnes of ^{235}U, in about 16 tonnes of HEU.

Naval spent fuel was sent to the Idaho National Engineering Laboratory for chemical reprocessing. About 9 tonnes of ^{235}U were recovered from this fuel until 1984, and a few more tonnes may have been recovered since then.

Recovered HEU from the Idaho facility was used to fuel the Savannah River production reactors, but this practice has been suspended. The Department of Energy has not decided on any other use for recovered HEU. Because of a lack of need for the recovered material, the separation plant at Idaho has been ordered permanently closed. Naval spent fuel will be stored and eventually disposed in a geological repository.

---

[7] See Cochran *et al.* (note 1), p. 71.

[8] See Cochran *et al.* (note 1), pp. 71 and 185.

[9] The lower bound results from assuming that all the cores contained 80 kg of weapon-grade uranium, and the upper bound assumes that the last 400 cores contained 200 kg.

[10] See Cochran *et al.* (note 1) and testimony of Admiral DeMars (note 5).

[11] See testimony of Admiral DeMars (note 5), p. 863.

Table 4.3. Estimated US naval fuel requirements up to the end of 1990

	Tonnes
Production of 97% enriched uranium	
Loaded into 600 cores	72 ± 14
Pipeline	24 ± 4
Stored (non-specification material)	7
Total	**103 ± 18**
Disposition of core material	
Fuel loaded in operating reactors (not including burn-up)	35
Spent fuel discharged (including burn-up)	16
Estimated fresh cores (in storage)	10
Total	**61**

Because naval cores take so long to fabricate, large quantities of weapon-grade uranium are located in the fuel manufacturing 'pipeline'. If the average amount of time to fabricate fuel were five to seven years, and the annual requirement for fuel were 4 tonnes of weapon-grade uranium, about 20–28 tonnes of unirradiated weapon-grade uranium are tied up in the pipeline at a given time. About 7 tonnes of weapon-grade uranium were produced for the Navy, but the material was not of the right quality for naval fuel. It is currently stored. Table 4.3 summarizes the above discussion.

IV. The former Soviet Union

Few historical enrichment production data are available for the former Soviet Union. Although a considerable amount of information about its enrichment facilities has been released in the past few years, most of it concerns the current programme.

As part of an offer to the United States to stop fissile material production for weapons, the Soviet Union announced in October 1989 that 'this year it is ceasing the production of HEU'.[12] The Soviet Union actually stopped making HEU for weapons two years earlier,[13] but it is unclear whether HEU production continued for non-weapon purposes, or was completely halted.

Besides freeing a considerable amount of enrichment capacity for civil uses, including exports, this step represented the end of a massive effort to make weapon-grade uranium that began immediately following World War II.

---

[12] Petrovsky, V. F., Deputy Head of the USSR Delegation to the 44th UN General Assembly, 'Statement on the item entitled Report of the IAEA', 25 Oct. 1989.

[13] NUEXCO, 'Conversion and Enrichment in the Soviet Union', *NUEXCO Monthly Report*, No. 272 (Apr. 1991); and Notes by Frank von Hippel on a meeting with Evgeny Mikerin, Deputy Minister for nuclear fuel cycle and uranium-isotope enrichment, Soviet Ministry of Atomic Power and Industry, 28 June 1991.

The Soviet enrichment complex

Immediately after World War II, the Soviet Union investigated several enrichment technologies, focusing particularly on gaseous diffusion and gas centrifuges.

Gaseous diffusion

The first process developed on an industrial scale used gaseous diffusion technology. The first industrial-scale plant entered service in 1949 at Verkhniy-Neyvinskiy near Sverdlovsk.[14]

Although the Soviet Government did not release information about the plant's capacity, some indication of its initial capacity is provided by US intelligence information. During 1949 and 1950, the USSR is believed to have ordered about 110 000 square metres of barrier material.[15] Since roughly 1 square metre is required for each installed separative work unit, the plant thus is estimated to have had a capacity initially of at least 100 000 SWU per year.

The first plant was relatively inefficient, and required additional development work.[16] Major problems were encountered in stopping corrosion by uranium hexafluoride gas and in developing the compressors which raise the pressure of the gas to the required level. The inefficiency of these compressors was a principal motivation for the development of gas centrifuges in the early 1950s.[17]

Information about other plants is sketchy, although it is known that more were built. It was reported in 1989 that a total of five gaseous diffusion plants were eventually built.[18] The capacities and dates of initial operation of other plants are unknown. By the mid- to late 1980s, these plants were almost completely phased out. In 1991, Evgeny Mikerin, then fuel-cycle director at the Ministry of Atomic Power (MAPI), said that uranium enrichment by gaseous diffusion had been abandoned.[19]

Gas centrifuges

In parallel to the gaseous diffusion programme, the USSR also developed gas centrifuges. Much of this development work was carried out by German scientists, who were taken to the USSR at the end of World War II. Based on their

[14] Chernov, A., 'Uranium enrichment in the USSR', paper presented at the Annual Symposium of the Uranium Institute, London, 6–8 Sep. 1989.

[15] Cochran, T., Arkin, W., Norris, R. and Sands, J., *Nuclear Weapons Databook, Vol. IV: Soviet Nuclear Weapons* (Ballinger: New York, 1989), p. 90, footnotes 127 and 128, citing CIA, Joint Atomic Energy Intelligence Committee, *Status of Soviet Atomic Energy Program*, CIA/SCI-2/50, 4 July 1950, p. 3 of accompanying 'Facts and Discussion'.

[16] See NUEXCO (note 13).

[17] See NUEXCO (note 13).

[18] 'Fact Sheet: Kyshtym complex and Soviet nuclear materials production', Natural Resources Defense Council, Washington, DC, July 1989.

[19] Hibbs, M., 'MAPI official says all four Soviet SWU plants are in Russian Republic', *Nuclear Fuel*, 11 Nov. 1991. In late January 1992, the portions of MAPI that concern the interests of the Russian Federation were absorbed into the newly created Russian Ministry for Atomic Energy (Minatom RF).

statements after they returned to the West, some information about the early centrifuge effort is available. From 1946 until 1953, the German research teams developed both a supercritical and a sub-critical machine.[20] The scientists were not informed about any production facilities built, and returned to Germany in the mid-1950s.

Based on recent information, the first gas centrifuge pilot plant went into operation at Verkhniy-Neyvinskiy in October 1957, and contained 2500 centrifuges.[21] It was followed by an industrial-scale plant in 1959 that contained 'several tens of thousands of centrifuges.'[22]

A full-scale plant with three modules was built there between 1962 and 1964.[23] It contained several hundred thousand sub-critical centrifuges, with rotors made out of aluminium alloy.[24] Sub-critical machines typically have a capacity of 2–3 SWU per year each, implying that the plant could have had a capacity in 1964 of the order of 1 million SWU per year.

There are some unconfirmed indications that gas centrifuges were becoming the dominant enrichment method by about 1970. Gernot Zippe, one of the principal captured German scientists who developed the Soviet centrifuges, said in 1971 that he noted talk among some enrichment specialists in Europe that 'the Russians already do a major share of their enrichment via centrifuges'.[25]

Enrichment capacity in 1992

In 1992 there are currently four enrichment plants, all of which use gas centrifuges and are in the Russian Federation.[26] They are located at the Electrochemistry Combine at Krasnoyarsk; the Electrolyzing Chemical Combine at Angarsk, about 50 km north-west of the Siberian city of Irkutsk; the Siberian Chemical Combine, near Tomsk; and at the Ural Electrochemistry Combine at Verkhniy-Neyvinskiy. Each of these four sites also had gaseous diffusion plants.

According to a Western uranium broker who visited the plant near Sverdlovsk, the capacity of this plant is about 2–3 million SWU a year. He said that this plant has produced the low-enriched uranium that has been exported to the West since the 1970s. A recent report, which cites an article by the former head of MAPI, says that the total enrichment capacity of these plants is about

---

[20] 'Background Note: research conducted in Soviet Union serves as basis for European centrifuge plants', *USSR Technology Update*, 31 Jan. 1989. The first one was a supercritical, flexible bellows-type machine, which was 3 m in length with a diameter of 58 mm and a peripheral speed of 240 m/s. The second units were sub-critical machines with peripheral speeds of up to 350 m/s, rigid rotors 58 and 100 mm in diameter, and rotor lengths of about 30 and 50 cm.

[21] See NUEXCO (note 13).

[22] See NUEXCO (note 13).

[23] See Chernov (note 14).

[24] See NUEXCO (note 13).

[25] Kolbenschlag, M., 'Zippe sure US also will go to "unbeatable" centrifuge enriching', *Nucleonics Week*, 19 Aug. 1971.

[26] 'Soviet Union: Soviets could sell 10-million SWU in mid-1990s', *Nuclear Fuel*, 29 Apr. 1991.

13 million SWU per year.[27] Other reports say that the total capacity is only 10 million SWU per year.

Up to 10 million SWU per year could be dedicated to the export market by the mid-1990s.[28] Most of this capacity was originally dedicated to military purposes, and has been freed up by recent cutbacks in the Soviet nuclear weapon programme.

Historical capacity

Little is known about the historical separative capacity of the enrichment plants. There is one estimate of the capacity during the late 1960s that has been widely misinterpreted in the open literature.

Many references in the late 1970s and 1980s cite the Soviet enrichment capacity as being 7–10 million SWU per year. This value appears to have originated from Manson Benedict, who published this estimate in the mid-1970s. He heard it while he was Chairman of the US General Advisory Committee in the late 1960s.[29] At the time, according to Benedict, this value was considered in a number of government circles as a reliable estimate of total Soviet capacity.

During the 1970s, total enrichment capacity is believed to have increased. According to a former senior US official, the USA estimated the capacity of the Soviet enrichment plants at about 12 million SWU per year during the 1970s. He did not remember an increase in the 1980s that would have 'warranted special concern'.

An indirect indication of the planned capacity of Soviet enrichment plants can be derived from recently available information about Soviet capacity to make uranium hexafluoride. In 1956, the USSR was able to increase production of uranium hexafluoride to between 2.5 and 8 tonnes an hour.[30] Assuming round-the-clock operation for 250 days a year, roughly 15 000–48 000 tonnes of uranium hexafluoride could have been produced per year. This is enough material to supply 'feedstock' to enrichment plants with a total capacity of between 13 and 43 million SWU per year, where we assume a 0.3 per cent tails assay. Because uranium conversion facilities can be built more quickly than enrichment plants, attaching any special significance to these estimates is bound to lead to a very substantial overestimate of Soviet enrichment capability, but the lower estimate of enrichment capacity can probably be interpreted as corresponding roughly to the minimum capacity planned in the Soviet Union.

[27] Novikov, V. and Lebedev, O., 'A Soviet view of the pros and cons of reprocessing', Uranium Institute Annual Symposium 1991, citing Konovalov, V., 'Uranium nuclear power plants and safety', *Pravda*, no. 125 (26573), 25 May 1991.

[28] See NUEXCO (note 13).

[29] Interview with Manson Benedict, Dec. 1986.

[30] See NUEXO (note 13).

Weapon-grade uranium inventory for weapons

Most non-governmental analysts have assumed that the Soviet stockpile dedicated to weapons is comparable to the US stockpile of roughly 500–600 tonnes. This estimate follows from multiplying the 32 000 weapons believed to be in the Soviet arsenal by an average of 15 kg per warhead, obtaining 480 tonnes of weapon-grade uranium in the weapons themselves. Another 100 tonnes is assumed to be in the production pipeline and in strategic reserves.

According to Mikerin, the Soviet Union has made 'well over' 500 tonnes of weapon-grade uranium, but he would not be more specific.[31] He also did not specify the fraction assigned to weapons and to strategic reserves.

Other Soviet officials and Western nuclear experts who have visited Soviet weapon facilities have provided anecdotal information about this inventory. They have given estimates that range from 500 to 600 tonnes to over 700 tonnes.

The US Government treats information about the Soviet inventory of fissile material as classified, but according to the 1989 congressional testimony of Troy Wade, then Acting Assistant Secretary, Defense Programs at the DOE, 'We currently estimate that the Soviet [weapon] stockpiles of both uranium and plutonium are comparable to or exceed those of the United States'.[32] Several DOE estimates of Soviet weapon-grade uranium are considerably higher than 700 tonnes.

Below, we attempt to estimate the Soviet inventory of weapon-grade uranium from the available information.

Estimates of cumulative production

Our estimates are based on the historical production information mentioned above. We have considered a low and a high case. In the low case we assume that production climbed linearly to about 7 million SWU per year in 1970, and then reached 10 million SWU per year in 1987. In the high case, the total capacity increased faster, rising to 10 million SWU per year in 1970 and to 12 million SWU per year in 1980, where it remained. The high case assumes a high capacity factor for the enrichment plants, a rapid installation of gas centrifuge plants, and competence on the part of the USSR. The low case assumes much slower production.

These assumptions lead to a total estimated production of 215–294 million SWU up to the end of 1987. Table 4.4 summarizes these results.

---

[31] Hibbs (note 19).

[32] Statement of Troy Wade before the Committee on Foreign Affairs and its Subcommittee on Arms Control, International Security and Science, in *International Plutonium Control Act–H.R. 2403*, Hearing before the Committee on Foreign Affairs and its Subcommittee on Arms Control, International Security and Science, US House of Representatives, 20 June 1989, 101st Congress (US Government Printing Office: Washington, DC, 1990), p. 29.

Table 4.4. Low and high estimates of Soviet enriched uranium production, 1950–87[a]

End year	Separative capacity (million SWU[b]/y.)		Cumulative (million SWU[b])	
	Low	High	Low	High
1950	0.1	0.2	0.1	0.2
1970	7	10	70	100
1980	8.8	12	149	210
1987	10	12	215	294

[a] Weapon-grade production halted in 1987.

[b] Kilogram separative work units.

Non-weapon requirements

A large fraction of this total would have gone to produce enriched uranium for various non-weapon uses:

1. About 40 million SWU were exported as low-enriched uranium to Western power reactors. (This represents deliveries up to the end of 1988, which we assume were all enriched before the end of 1987.)

2. Domestic, East European and Finnish power reactors required roughly 25 million SWU to the end of 1987.

3. Soviet plutonium production reactors probably used slightly enriched uranium. If the Soviet reactors used the same type of fuel-cycle as the US Hanford production reactors, the enrichment of the fuel would have required about 5 million SWU.

4. The Soviet Union is reported to have two or three reactors dedicated to tritium production. Without specific information, we assume that up to the end of 1987 the Soviet Union produced about 200 kg of tritium (not corrected for decay). This is similar to the amount estimated as produced by the USA.[33] The type of reactor used to produce tritium is unknown, so we roughly estimate SWU requirements in two different ways.

(a) We first assume that the reactors are similar to the N-reactor at Hanford (water-cooled, graphite-moderated, 1.2 per cent fuel). This reactor could produce about 3 kg of tritium a year.[34] At this rate, 66 reactor-years of operation would be required to make 200 kg of tritium. Each year, the N-reactor required about 140 000 SWU, where the spent fuel is processed and the recovered uranium is re-enriched.[35] Production of 200 grams of tritium would therefore require about 9 million SWU.

[33] See Cochran *et al.* (note 1).

[34] US House of Representatives, *Department of Energy Authorization Legislation (National Security Programs) for Fiscal Year 1982*, Hearing before the Subcommittee on Procurement and Military Nuclear Systems, Committee on Armed Services, 97th Congress, 1st session (US Government Printing Office: Washington, DC, 1981), p. 171.

[35] See Cochran *et al.* (note 1).

(*b*) Next, we assume that the reactors are similar to the Savannah River reactors (heavy water-moderated and -cooled). The NRDC estimated that the production of 1 kg of tritium in a Savannah River reactor would require the consumption of about 96 kg of ^{235}U.[36] Assuming that the fuel is 60–90 per cent enriched, and about 40 per cent of the initial ^{235}U is consumed, then about 10 million SWU would have been required. This estimate assumes no recycling of the spent fuel, which is unlikely. If up to half the separative work was saved through recycling, then as little as 5 million SWU would have been required. As can be seen, this estimate is highly uncertain, but the above shows that the requirement for tritium production is unlikely to have exceeded 10 million SWU.

We assign 7 million SWU to the production of tritium.

5. Some 325 reactors are on about half that number of Soviet naval vessels as of 1990.[37] We assume that enriched uranium enough for four times that number of reactor cores had been produced by the end of 1987. Soviet naval reactors are reported to use 10 per cent enriched uranium, although more recent reports have stated that the enrichment varied up to weapon-grade.[38] In any case, we will assume that each reactor requires between 40 and 70 kg of ^{235}U, requiring between 52 and 91 tonnes of ^{235}U to produce all the cores. If the fuel was 10 per cent enriched, its production would have required about 10–18 million SWU. If the fuel was all fully enriched, it would have required about 12–20 million SWU. We pick the mid-point of these ranges, or roughly 15 million SWU.

6. A total of over 30 research reactors have generated an estimated 3 million MWd of energy up to the end of 1987.[39] Generation of this amount of heat would require roughly 10 tonnes of 90 per cent enriched fuel. Soviet space reactors are estimated to have required about 1 tonne of weapon-grade uranium. This amount of fuel would require about 2 million SWU. Exports of enriched uranium fuel for foreign research reactors would not add significantly to this total.

7. About 10 tonnes of weapon-grade uranium are estimated to have been consumed in nuclear weapon tests, or the equivalent of two million SWU.

8. Process losses might have consumed about 3 per cent of the amount of enriched uranium produced. This would entail the loss of an equivalent of 8 million SWU in the low and high estimate respectively.

9. The uranium in process at the end of 1987 is estimated to be no more than 10 per cent of the annual separative work production, equivalent to about

[36] See Cochran *et al.* (note 1), p. 185.

[37] Nuclear Notebook, 'Nuclear weapons at sea, 1990', *Bulletin of the Atomic Scientists*, Sep. 1991.

[38] See 'Fact Sheet' (note 18); and Handler, J., 'Preliminary report on Greenpeace visit to Vladivostok and areas around the Chazma Bay and Bolshoi Kamen submarine repair and refuelling facilities, 9–19 October 1991', Greenpeace, Washington, DC, 6 Nov. 1991.

[39] IAEA, *Nuclear Research Reactors in the World* (IAEA: Vienna, 1991), table 14. We assume an average capacity factor of 70% for Soviet research reactors.

Table 4.5. Estimated Soviet consumption of separative work units and equivalent production of weapon-grade uranium up to the end of 1987

	SWU (x 10^6)	WGU equivalent (tonnes)[a]
LEU[b] exports	40	200
LEU for domestic, East European and Finnish power reactors	25	125
Fuel for production reactors, plutonium production	5	25
Tritium production	7	35
Fuel for naval propulsion reactors	15	75
Fuel for domestic and foreign research	2	10
Consumption in nuclear weapons tests	2	10
Process losses (3% of original enrichment output)	8	40
In process at enrichment plants	1	5
Working inventory	5	25
Total	**110**	**550**

[a] These numbers represent the amount of weapon-grade (WGU) that could have been produced with the enrichment work used to produce this particular enriched product, assuming a 0.3% tails assay.
[b] Low-enriched uranium.

1 million SWU. In addition, a working inventory of several months' worth of production, roughly 5 million SWU, might be needed in the enrichment complex in order to meet fluctuations in demand and provide a stock of various enrichment levels.

Table 4.5 summarizes the above draw-downs. In total, about 110 million SWU would have gone to non-weapon uses. We attach an error margin of plus or minus 25 per cent.

Subtracting these requirements from the total enrichment effort leaves about 105–184 million SWU. This is equivalent to about 520–920 tonnes of weapon-grade uranium, at a tails assay of 0.3 per cent. This estimate has a large amount of uncertainty attached to it.

V. China

China first produced weapon-grade uranium in early 1964 at the Lanzhou uranium enrichment plant, located in Gansu province (central China). During the next several months, the plant produced enough material for China's first atomic test in October 1964.

During the late 1970s and early 1980s, China redirected its nuclear infrastructure to serve civilian goals. China is reported to have stopped producing weapon-grade uranium for weapons in 1987.[40]

[40] MacLachlan, A. and Hibbs, M., 'China stops production of military fuel: all SWU capacity now for civil use', *Nuclear Fuel*, 13 Nov. 1989. The 1987 date is from a personal communication from M. Hibbs, who was told this by the head of the China Nuclear Energy Industry Corporation.

Gaseous diffusion plants

As a result of Soviet–Chinese collaboration in the 1950s, the Soviet Union assisted China in the construction of a gaseous diffusion enrichment plant.[41] Construction started in 1957, and by the time the Soviet experts left the plant site in August 1960 China had received a significant amount of the necessary technology and components.[42] This included the diffusers (with separation membranes) and much of the specialized equipment necessary to monitor and run the plant.

In mid-January 1964, the plant began producing weapon-grade uranium. The capacity of the plant at that time is unknown, although we estimate that nominal capacity was between 50 000 and 100 000 SWU per year. At this capacity, the plant could produce about 250–500 kg of weapon-grade uranium each year.

The Chinese encountered problems in learning to use and duplicate the Soviet-supplied components and in manufacturing missing equipment. This delayed both the opening of the plant and its expansion.

The first increase in capacity occurred in the mid-1970s, when the plant was renovated. *Nucleonics Week* reported in 1978 that the plant's capacity was 180 000 SWU per year.[43]

A second period of expansion of the plant occurred during the early 1980s, when China achieved an 'enormous breakthrough in separation membrane technology'.[44] This development led to a further increase in separation efficiency. These improvements during the 1970s and 1980s resulted in a 'many-fold' increase in production capacity over initial output.[45] Currently, the plant is believed to have a capacity of about 300 000 SWU per year.[46]

According to an official at NUEXCO, a large uranium broker, China is building a new cascade at Lanzhou to produce low-enriched uranium that will be more suitable for export, but this facility is not expected to be completed until the mid-1990s. Chinese low-enriched uranium contains traces of ^{232}U and ^{236}U that were deposited into enrichment cascades by re-enriching production reactor fuel. Western fuel fabricators are not inclined to buy it, particularly with the current glut of commercial enriched uranium.

China is reported to have a second gaseous diffusion plant, although details of it are sketchy, including its location and its operational status. *Nuclear Fuel* reported, citing Western sources, that the capacity of the second plant is a little larger than that of the first plant.[47] One Western official said that the plant is in the south of China and has operated for at least 10 years. If this last information is correct, the plant might have been built as a 'third-line' weapon manufactur-

---

[41] The description of the Chinese development of gaseous diffusion plants is based on *China Today: Nuclear Industry*, Apr. 1987. Selections translated by the Foreign Broadcast Information Service, JPRS-CST-88-002, 15 Jan. 1988; and JPRS-CST-88-008, Washington, DC, 26 Apr. 1988.

[42] *China Today* (note 41).

[43] Mainland China talking to French, Germans, about nuclear power,' *Nucleonics Week*, 12 Jan. 1978.

[44] See *China Today* (note 41), p. 14, in translation.

[45] See *China Today* (note 41), p. 15, in translation.

[46] See MacLachlan and Hibbs (note 40).

[47] See MacLachlan and Hibbs (note 40).

ing site. During the Cultural Revolution, the Chinese leadership ordered the construction of strategic facilities in the interior of the country, far from the more vulnerable coasts and borders. In the case of plutonium production, this pattern was also followed (see chapter 3).

Gas centrifuge plants

China has conducted R&D on gas centrifuges since 1958. Chinese officials have stated that China might build a gas centrifuge or laser enrichment plant to replace the gaseous diffusion plants.[48] Recently, Russian officials said that China had approached Russia about buying a 200 000 SWU per year gas centrifuge plant to produce low-enriched uranium for domestic nuclear power requirements.[49]

China is suspected in the West of having gained access to Western gas centrifuge technology and equipment indirectly through Pakistan. Details of this exchange remain sketchy, however.

Inventory of weapon-grade uranium

Historical production

China's inventory of weapon-grade uranium can be roughly estimated from the above information. We assume that production in the first plant averaged about 75 000 SWU per year from 1964 until the end of 1977, at which time the capacity was increased to 180 000 SWU per year. During an expansion which we assume started in 1980 and lasted until 1985, the plant's capacity linearly increased to 300 000 SWU per year. Under these assumptions the first gaseous diffusion plant is estimated to have produced about 3.3 million SWU up to and including 1987.

The output of the second plant is highly uncertain. We assume that it started operating in 1975 at a low capacity and increased linearly to 300 000 SWU per year in 1987. Total output of this plant is estimated at 1.8 million SWU.

Since uncertainties about output at both plants are high, we estimate total production as a range of 3–5 million SWU.

Non-weapon uses

The enrichment programme would have supplied about 1 million SWU of enriched uranium for several non-weapon purposes, primarily fuel for research, production and naval reactors, and for nuclear weapon tests:

    1. China's research reactors would have required up to 230 000 SWU.[50]

---

[48] See MacLachlan and Hibbs (note 40).

[49] Hibbs, M., 'China wants 200,000 SWU/yr centrifuge plant from Russia', *Nuclear Fuel*, 26 Oct. 1992.

[50] See IAEA (note 39). Combined, the research reactors could have produced about 380 000 MWd of heat, assuming a 70% capacity figure. Two-thirds of this heat was produced by the 125 MWth reactor at

2. Production of low-enriched uranium for five nuclear-powered submarines and one land-based prototype would have required roughly 115 000 SWU.[51]

3. About 700 kg of weapon-grade uranium, or about 140 000 SWU, would have been consumed in about 36 nuclear weapon tests, where we assume an average of 20 kg per test.

4. China's graphite-moderated, water-cooled production reactors are believed to use a fuel cycle similar to US production reactors. If China produced 2.5 tonnes of weapon-grade plutonium, fuel for these reactors would have required about 100 000 SWU.[52]

5. Process losses of 3 per cent would have consumed about 150 000 SWU. A working inventory, equivalent to five months' supply, would have required another 250 000 SWU.

Inventory for weapons

Subtracting these requirements, we are left with about 2–4 million SWU, or enough for between 10 and 20 tonnes of weapon-grade uranium. This number is very uncertain.

China is estimated to have stockpiled roughly 300 weapons, although this number could be two to three times larger.[53] Its weapons are believed to depend heavily on weapon-grade uranium, so we estimate that each weapon contains on average 20 kg of weapon-grade uranium. China would therefore need about 6 tonnes of weapon-grade uranium to build this number of weapons. If the arsenal is two to three times greater, then 12–18 tonnes of weapon-grade uranium would be needed.

VI. The United Kingdom

The UK has obtained weapon-grade uranium for its nuclear arsenal from a gaseous diffusion enrichment facility at Capenhurst and from the USA. Little is known about the amount produced at Capenhurst or acquired from the USA.

The high-enrichment section of the Capenhurst plant closed in 1963. The reason for closing appears to have been related to the high cost of enriching uranium in the UK and the appearance of an alternative supply of weapon-grade uranium.[54]

---

the Reactor Operation Institute that uses weapon-grade uranium fuel. Assuming that about 40% of the uranium-235 is fissioned or converted into uranium-236, fuel for these reactors would require less than 230 000 SWU.

[51] According to *China Today*, Part II (note 41), Chinese submarine reactors use low-enriched uranium fuel. We assume that a typical naval core contains about 800 kg of 5% enriched uranium fuel (40 kg of uranium-235). Assuming that a total of 20 cores have been fabricated, total requirements would be 16 000 kg of 5% enriched uranium, or 115 000 SWU.

[52] See Cochran *et al.* (note 1).

[53] Fieldhouse, R. W., 'Chinese nuclear weapons: a current and historical overview', NWD-91-1 (Natural Resources Defense Council: Washington, DC, Mar. 1991).

[54] Simpson, J., telephone interview, Feb. 1992.

The UK signed a Mutual Defence Agreement with the USA in the late 1950s (see chapter 3). Under this agreement, the UK traded a growing surplus of plutonium for weapon-grade uranium and tritium from the USA. Later, it bought weapon-grade uranium directly from the USA.

Weapon-grade uranium supply

Capenhurst

Britain built the gaseous diffusion plant at Capenhurst in the 1950s. The plant started operating between 1954 and 1957, first producing weapon-grade uranium in 1957. The nominal capacity of the plant is reported to have been between 0.4 and 0.6 million SWU per year.[55] This corresponds to a maximum production of 2–3 tonnes of weapon-grade uranium per year, at a tails assay of 0.3 per cent, but some of this capacity was dedicated to other purposes, including re-enriching recovered uranium from plutonium production reactor fuel.

Although no evidence exists that the plant experienced start-up problems, it is unlikely that the plant would have operated at peak production during its initial years of operation. As a first approximation, we estimate that an average of 0.2 million SWU a year was dedicated to weapon-grade uranium production from 1957 until 1963. Total output is estimated at 1.2 million SWU, enough for 6 tonnes of weapon-grade uranium.

Britain decided in the late 1970s to build a military centrifuge plant at Capenhurst able to produce HEU. It started operating in 1984, but it is believed that weapon-grade material was not produced.

Barter agreement with the USA

Estimates of the amount of weapon-grade uranium transferred to Britain under the Mutual Defence Agreement vary between 5 and 9 tonnes.[56] The discrepancy results from not knowing the 'exchange value' of the 4 tonnes of fuel-grade plutonium and what may have been 1 tonne of weapon-grade plutonium that Britain sent to the USA (see chapter 3). The plutonium was exchanged for an equal amount of weapon-grade uranium or for 1.76 times as much. Simpson has suggested that the weapon-grade plutonium was traded at the higher value and the fuel- or reactor-grade plutonium at the lower value.[57] This implies that the British plutonium was worth about 6 tonnes of weapon-grade uranium. A small fraction of the plutonium was traded for tritium, but its contribution is ignored here.

Britain obtained additional weapon-grade uranium in the 1980s from the USA. It contracted for up to 100 000 SWU per year from 1981 until 1986 for

[55] Benedict, M., Pigford, T. and Levi, H., *Nuclear Chemical Engineering* (McGraw-Hill: New York, 1981), p. 816.

[56] See Cochran *et al.* (note 1), p. 190.

[57] Simpson (note 54).

military purposes.[58] We have no information about the amount of separative work actually used, or the purpose of this relatively large contract. But one objective was to obtain weapon-grade uranium for submarine reactors.[59]

In 1987, the UK contracted for a one-off shipment of an unspecified amount of weapon-grade uranium from the USA.[60] The material was reported to be in metallic form and was taken from the US weapons stockpile that had been produced before 1964. The purpose of the weapon-grade uranium was not specified, although the fact that it was metal might imply it was for British weapons or naval reactors.

We can only estimate very roughly the total amount of weapon-grade uranium imported from the USA under these last two transfers. The most straightforward assessment is that only a fraction of the five-year enrichment contract was for producing weapon-grade uranium, and this material was dedicated to naval reactors. The one-time shipment is assumed to be for weapons, and we estimate that about 1 tonne was acquired.

Demand for weapon-grade uranium

Britain had several non-weapon requirements for weapon-grade uranium:

1. Weapon-grade uranium is used to fuel the UK's nuclear-powered submarines, which numbered 20 in 1990. Some of this material would have been taken from its inventory, but not all of it. The British Government announced in 1982 that it needed to obtain weapon-grade uranium for its submarines by enriching it to an intermediate level at Capenhurst and then sending it to the USA for further enrichment to weapon-grade.[61] We do not know when the UK first received this enriched uranium from the USA, but we estimate that the UK had to fabricate a total of about 50 cores before it arrived.[62] We estimate that each core required about 40 kg of weapon-grade uranium, requiring 2 tonnes in total.

2. Civil research reactors might have required some weapon-grade uranium from the military inventory. We believe that this amount was small, and we ignore it here. The main reason is that in addition to receiving military shipments of weapon-grade uranium from the USA, the UK also received over 2 tonnes of HEU from the USA for civil purposes. Some of the imported material was used in British civil reactors, and some of it was re-exported after fabrication into fuel. The largest research reactor using HEU, the 65 MW Dounreay Fast Reactor (DFR) which operated from 1959 until 1977, did not use weapon-

---

[58] *Nuclear Fuel*, 13 Apr. 1981, p. 9.

[59] John Nott, Written Parliamentary Answer, *Parliamentary Debate, House of Commons [Hansard]*, 23 June 1982, cols 128–29; and John Lee, Written Parliamentary Answer, *Hansard*, 17 July 1985, cols 181–82.

[60] Associated Press, 'Oak Ridge uranium headed for Britain, newspaper reports', 28 Nov. 1987.

[61] Nott (note 59).

[62] Fishlock, D., 'New design for submarine reactors', *Financial Times*, 12 June 1981. He wrote that prior to 1982, about 40 cores were fabricated. We estimate that another 10 were fabricated before the new material arrived from the USA.

grade uranium fuel. The reactor started with 46 per cent enriched uranium, and then used 75 per cent enriched uranium.[63] When the reactor started, HEU was in short supply. This shortage apparently led to an early decision to reprocess the spent fuel, lowering further the amount of enriched uranium dedicated to this reactor.[64]

The UK was also constrained in using HEU from military stockpiles in its research reactors after it joined the European Community in 1972. Thereafter, it had to submit its civil facilities to Euratom safeguards.

3. By the end of 1990, the UK's 43 nuclear tests would have required about 650 kg of weapon-grade uranium.

These draw-downs of weapon-grade uranium total about 2.7 tonnes. Subtracting them from the above estimate of 13 tonnes leaves a stockpile of about 10 tonnes of weapon-grade uranium dedicated to weapons, the pipeline and strategic inventories. We attach an uncertainty of plus or minus 5 tonnes to our estimate.

Britain is estimated to have between 200 and 300 nuclear weapons. If each weapon has about 15 kg of weapon-grade uranium, about 3–4.5 tonnes would be in those weapons. If Britain increases the number of warheads to 500 when it deploys the Trident submarines, another 3 tonnes of weapon-grade uranium would be required. Building these new warheads would also require a large amount of weapon-grade material in the pipeline, perhaps as much as a few tonnes of material.

VII. France

Least is known about France. It started producing weapon-grade uranium in 1967 or 1968 at the Pierrelatte gaseous diffusion plant. The material was specifically produced for thermonuclear weapons, which it first tested in 1968. We do not know if France has stopped making weapon-grade uranium for weapons although, with the end of the cold war, it is considering a halt to production.

The nominal capacity of this plant has been reported as between 0.3 and 0.6 million SWU per year.

Assuming that an average of 0.2 million SWU a year were dedicated to producing weapon-grade uranium from the end of 1970 to the end of 1991, France would have produced about 20 tonnes of weapon-grade uranium. On the basis of the available information about the Pierrelatte enrichment plant, we are unable to verify that this amount of weapon-grade uranium was produced. However, we believe that this estimate is an upper bound on production.

France is estimated to have 500–600 nuclear weapons. At 15 kg per weapon, France would require between 7.5 and 9 tonnes of weapon-grade uranium. It

---

[63] 'The Dounreay fuel cycle facilities of AEA Fuel Services', *Atom*, Jan. 1992. Originally published under 'Flexible reprocessing at Dounreay', *Nuclear Engineering International*, Nov. 1991.
[64] See 'The Dounreay fuel cycle' (note 63).

would have also consumed about 2.8 tonnes of weapon-grade uranium in 183 nuclear weapon tests to the end of 1990. Its submarines are believed to use low-enriched uranium fuel, and therefore would not change this estimate. We therefore assign France an estimate of roughly 10–20 tonnes of weapon-grade uranium.

Part III
Principal civil inventories

Part III covers the separation and use of plutonium and HEU in civil nuclear programmes. Chapters 5 to 7 trace the production, storage, separation and use of all the plutonium which has been produced at civil power reactors. This covers the majority of the world's plutonium. A proportion of this material has been separated, and a proportion of this has been recycled in fast and thermal reactors. In the coming two decades larger amounts of plutonium are expected to be separated.

Defining what is a power reactor is straightforward in a non-nuclear weapon state. In a nuclear weapon state less so since plutonium discharged from power reactors has on occasion been taken into military stockpiles, while military plutonium production reactors have been operated to produce electricity. Here we include only those reactors which are identified as having been built primarily for electricity production. In this category we do not include reactors such as the N-reactor in the USA, or Calder Hall and Chapelcross in the UK.

Chapter 5 assesses the production and distribution of plutonium in commercial power reactors. It presents estimates of spent fuel discharges and plutonium arisings region-by-region and country-by-country, and considers possible trends to the year 2010.

Chapter 6 reviews the separation of plutonium from spent fuel at commercial and R&D reprocessing plants. It focuses on the eight main reprocessing countries: the UK, France, Japan, Germany, Belgium, the Soviet Union, the United States and India. It gives summary data for plutonium separation from power reactor fuel and addresses the important question of ownership of the material. Although most separation has occurred in nuclear weapon states, much of this material belongs to utility companies in non-nuclear weapon states.

Chapter 7 analyses plutonium use in R&D reactors and in commercial power reactors. Scenarios are presented of the possible future use of plutonium, primarily through thermal recycling. Consumption is compared with production, showing that large surplus stocks of plutonium may occur if the rate of reprocessing increases as expected.

Chapter 8 briefly reviews the production and use of HEU in civilian reactors. This HEU was largely produced in the military enrichment facilities of the nuclear weapon states. The United States, which supplied most of this HEU, had a policy until recently of taking back the irradiated HEU fuel for reprocessing. Afterwards, the recovered HEU re-entered its military programme. Since HEU is not a by-product of nuclear reactors, its use in the civil fuel cycle can be eliminated. To that end, many countries have taken steps to replace HEU fuels with low-enriched uranium fuels that cannot be used in nuclear weapons.

5. Plutonium produced in power reactors

I. Introduction

Plutonium production is the inevitable consequence of irradiating uranium in a nuclear reactor. It arises primarily through neutron capture by ^{238}U to form ^{239}U and then, via decay, ^{239}Pu (see chapter 2). Heavier isotopes of plutonium, from ^{240}Pu to ^{243}Pu, are produced by further neutron captures. Only ^{239}Pu and ^{241}Pu are fissile when irradiated with thermal neutrons and have value as fuel in today's power reactors.

The rate at which plutonium is produced in nuclear fuel depends on the reactor type and its characteristics (moderator, coolant and fuel type), the enrichment of the fuel, where the fuel was located in the reactor, and how the reactor is operated. In general, the longer fuel is irradiated, the greater the amount of plutonium contained within it. Fuel in commercial power reactors is typically irradiated for longer periods than fuel in military plutonium production reactors (see chapter 3). Increased irradiation times change the isotopic composition of the plutonium, with lower proportions of the fissile isotopes ^{239}Pu and ^{241}Pu in highly irradiated fuel. Plutonium in fuel discharged from a commercial reactor will typically have a fissile content of about 70–73 per cent (see table 2.1). Weapon-grade material usually contains more than 93 per cent ^{239}Pu.

Detailed information on reactor operation and fuelling is not usually made public. Moreover, even with such information, there will always be some uncertainty about the amount of plutonium produced in irradiated fuel. The British Central Electricity Generating Board claimed, for instance, that because of errors in its measurement of heat output it could not calculate plutonium production in a Magnox reactor in any one year to an accuracy of greater than plus or minus 5 per cent.[1] Only when a fuel element has been dissolved at a reprocessing plant can an accurate measurement (that is, with an error margin of below plus or minus 1 per cent) be made of its plutonium content.

In making estimates of plutonium production here, a number of assumptions have had to be made about fuelling strategy. Usually the only available hard evidence on reactor operation is the electrical output, although there is now a growing body of data on reactor spent fuel discharges. This can be used to calculate total heat output figures, from which plutonium production can be inferred. Heat output is calculated by dividing electrical output by the efficiency of the reactor (usually around 30 per cent). If a reactor is at equilibrium, fuel burn-up can then be inferred by dividing this heat production (measured in

[1] Hinkley Point 'C' Inquiry, UK, Transcript of evidence day 147, UK, 1989, pp. 19–25.

Megawatt days-thermal) by the amount of fuel discharged from the reactor (either by using direct information or by making an assumption about the rate of fuel discharge—usually between one-sixth and one-half of the core per annum). The fuel burn-up can in turn be used to derive plutonium inventories. For particular fuel enrichments, plutonium is generated according to well-known physical principles. A more detailed description of this methodology is provided in appendix B.

Estimates presented in this chapter closely follow the methods used by the Systems Studies Section of the Department of Safeguards at the IAEA in a series of reports produced during the early and mid-1980s.[2] For some specific countries such as the UK and the USA much better information is now available on the public record and this has been incorporated. In making forecasts here, attention has been given to the effects of improvements in fuel technology and higher fuel burn-ups.

Fuelling, burn-up and plutonium production histories have been modelled for each of the world's civil power reactors. In this chapter only summary data are presented, together with some impression of their reliability. Rather than producing annual plutonium discharge figures, the emphasis is on decadal estimates and forecasts of cumulative amounts from 1980 to 2010. The overall level of accuracy achieved is believed to be within plus or minus 10 per cent.

II. The fuel cycle in civil reactor systems

Nuclear fuel in reactors has to be replaced when its fissile content becomes depleted. In some gas-cooled, CANDU and RBMK reactors fuel may be replaced while the reactor is operating. This is known as 'on-load refuelling'. For most reactors, including the dominant LWR, each irradiation campaign is followed by a shut-down when a proportion of the core is replaced.

On discharge, the fuel contains uranium, including a residual amount of fissile ^{235}U, plutonium, and waste fission and transuranic products. It is hot and highly radioactive. Fuel is normally stored under water at the reactor and allowed to cool, for one year at Magnox stations and for at least three at LWRs. There are then two options. The fuel may be sent for reprocessing, or it can be stored (see figures 5.1 and 5.2). For Magnox reactor fuel, which is uranium metal with a magnesium oxide cladding, there is generally held to be no alternative to reprocessing. Ceramic oxide fuels with corrosion-resistant cladding can be safely stored for periods of decades.

If the fuel is reprocessed, the plutonium will be handled separately. If not, it will remain locked in the fuel matrix. Fuel left in store may either be treated as a waste, or held until some future time when reprocessing may be justified.

---

[2] Bilyk, A., *Forecast of Amounts of Plutonium at Power Reactors subject to Safeguards (1981–1990)*, STR-125 (Department of Safeguards, IAEA: Vienna, June 1982); Mal'ko, M., *Estimation of Plutonium Production in Light Water Reactors*, STR-226 (Department of Safeguards, IAEA: Vienna, Nov. 1986).

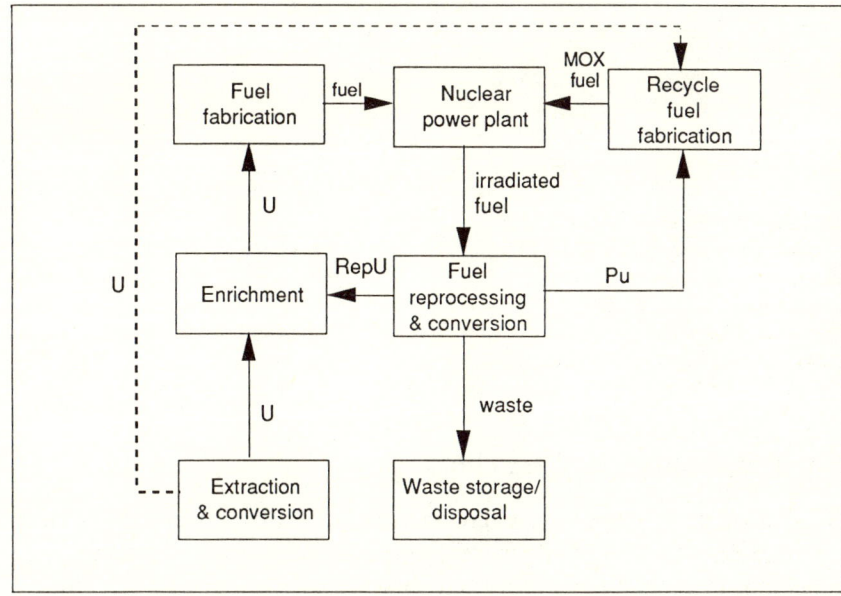

Figure 5.1. The nuclear fuel cycle including reprocessing

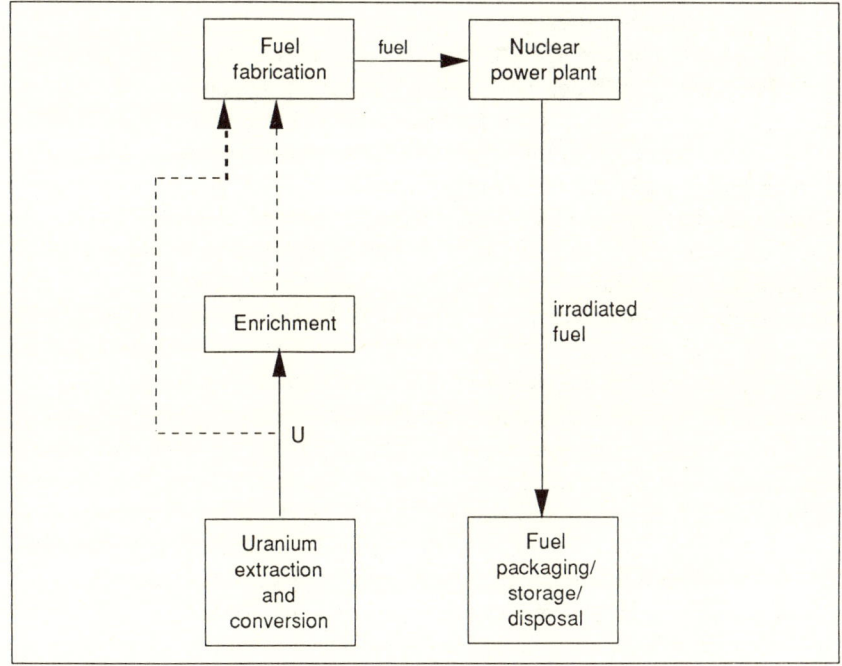

Figure 5.2. The once-through nuclear fuel cycle

Table 5.1. Fuel characteristics of power reactors

Reactor	Fuel type	Fuel cladding	Typical initial enrichment (% ^{235}U)
Magnox (GCR)	Metallic	Magnesium oxide	0.7
Advanced gas-cooled reactor (AGR)	Oxide	Stainless steel	2.1 – 2.5
Pressurized heavy-water reactor (CANDU)	Oxide	Zirconium	0.7
Pressurized water reactor (PWR)	Oxide	Zirconium	3.8
Pressurized water reactor (VVER)	Oxide	Zirconium	3.6
Boiling water reactor (BWR)	Oxide	Zirconium	3.2
Graphite-moderated light-water reactor (RBMK)	Oxide	Zirconium	2.4 – 3.0

III. Fuelling strategy and fuel burn-up

The fissile content of uranium is exploited differently across the range of thermal reactor types. The only thing that these reactor designs have in common is the slowing of neutron speeds to thermal velocities in the reactor core through their interaction with a moderator. This sets them apart from 'fast' reactors in which neutron velocities are not reduced. At slower velocities the likelihood of fission occurring in fissile atoms is enhanced (see chapter 2).

The main reactor systems treated here are listed in table 5.1 together with their typical fuel characteristics.

Commercialized nuclear reactor systems use either natural or enriched uranium. The chief difference between these reactor systems is their power density—the amount of heat produced per unit volume of the core. LWRs (fuelled with enriched uranium) typically have power densities five times greater than heavy water reactors (natural uranium) which, in turn, have higher power densities than gas-cooled Magnox reactors (natural uranium). LWRs are therefore more compact, with smaller fuel cores. A pressurized water reactor (PWR) with an electric power rating of 1 gigawatt (GWe) typically has a core weighing about 90 tonnes, whereas an equivalent Magnox station would have a core of about 1000 tonnes. Per unit of electricity produced, some 8–10 times as much fuel is discharged from a Magnox reactor as from an LWR.

Differences in power density have an important effect on plutonium production. Lightly irradiated fuel (such as is discharged from gas-cooled reactors) contains less plutonium, per kilogram, than more highly irradiated fuel. As discussed in chapter 2, it is also less contaminated with the higher-numbered isotopes of plutonium. The standard measure of fuel irradiation is burn-up. Fuel burn-up is the cumulative thermal power generated per unit weight of fuel (usually measured in megawatt-days per tonne of fuel, MWd/t), and can be used in determining how much plutonium is produced in the fuel. Fuel burn-up may also be expressed as a percentage of the initial fissile plutonium which has

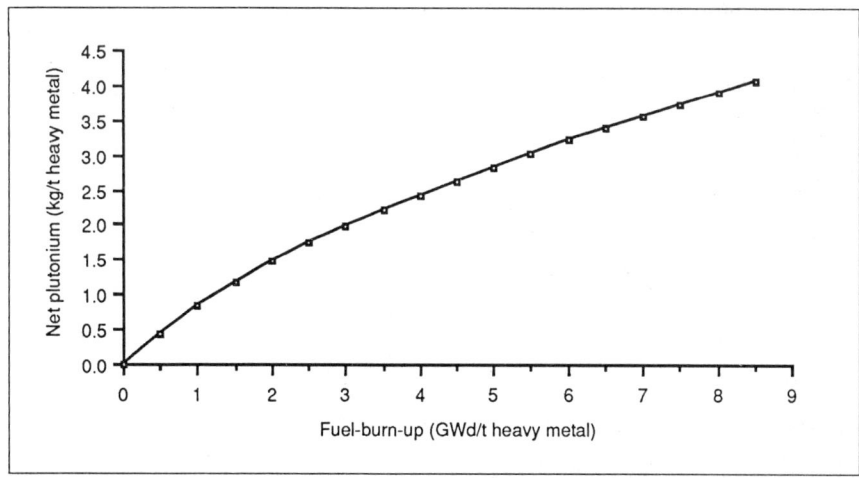

Figure 5.3. Specific plutonium production as a function of fuel burn-up, natural uranium fuel

Sources: Bilyk, A., International Atomic Energy Agency, *Forecast of Amounts of Plutonium at Power Reactors Subject to Safeguards (1981–1990)*, STR-125 (IAEA Department of Safeguards: Vienna, June 1982); Turner, S. E., *et al.*, *Criticality Studies of Graphite-Moderated Production Reactors*, Report prepared for the US Arms Control and Disarmament Agency, SSA-125 (Southern Sciences Applications Inc., Washington, DC, Jan. 1980).

been fissioned. Graphs showing the relationship between fuel burn-up and plutonium content in natural and enriched uranium spent fuel are given in figures 5.3 and 5.4.

Natural uranium-fuelled reactors generally operate at lower burn-ups than those fuelled with enriched uranium: magnox fuel burn-ups typically range from 3000 to 5000 MWd/t; CANDU burn-ups from 6000 to 8000 MWd/t; and AGRs, fuelled with slightly enriched fuel, from 18 000 to 21 000 MWd/t. Fuel in a typical PWR, is enriched to 3.25 per cent ^{235}U and irradiated to 30 000 MWd/t during the course of three annual campaigns (one-third of the reactor core being replaced each year). Boiling water reactor (BWR) fuel typically has an initial enrichment of 2.6–2.8 per cent and achieves burn-ups of 28 000 MWd/t.

In recent years PWR and BWR operating experience and improved fuel technology have led utilities to increase fuel burn-ups. This may be achieved in two ways: by increasing the length of time between refuelling; or by reducing the number of assemblies replaced during each campaign (one-quarter, one-fifth, or one-sixth core, rather than one-third). Initial enrichments have to be raised in each case, although by slightly less in the second option. Depending on the approach followed, PWR burn-ups may reach 60 000 MWd/t by the end of the century, while for BWRs they could increase to 40 000 MWd/t.

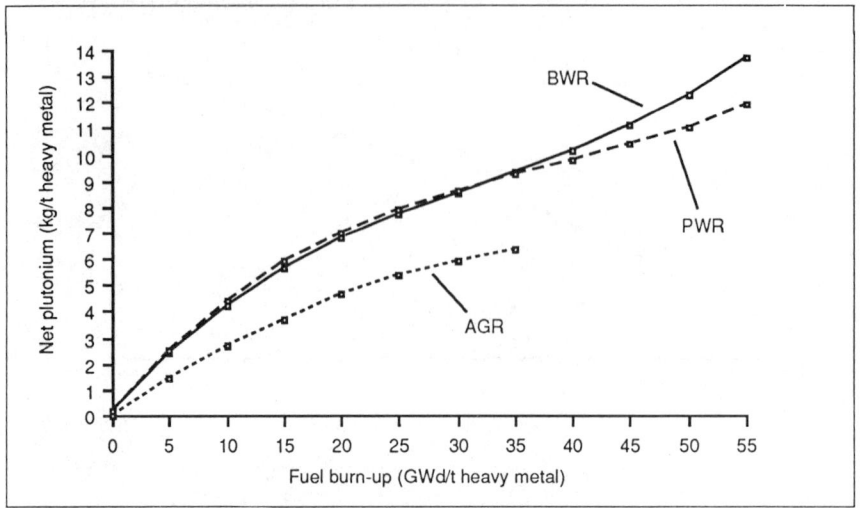

Figure 5.4. Specific plutonium production as a function of fuel burn-up, enriched uranium fuel

Note: BWR—boiling water reactor; PWR—pressurized water reactor; AGR—advanced gas-cooled reactor.

Source: Mal'ko, M., International Atomic Energy Agency, *Estimation of Plutonium Production in Light Water Reactors*, STR-226 (IAEA Department of Safeguards: Vienna, Nov. 1986).

IV. A sketch of methods

Information on reactor capacities, core weights, nominal thermal efficiencies and annual electrical outputs is widely available for most of the world's reactors.[3] Using these data, assumptions can be made about fuelling strategies to give estimates of the amount of fuel discharged by a reactor. In general, it is assumed here that reactor operators have sought to optimize their use of fuel over the long term. Taking all this together, average fuel burn-ups can be calculated. By using the conversion factors shown in figures 5.3 and 5.4, plutonium production can then be deduced.

Depending on the information available, a variety of methods have been used to model plutonium production in the world's civil reactors.

1. The best situation is where accurate fuel discharge figures are available. Burn-up is calculated using the gross electrical output and the thermal efficiency of each reactor. Only the UK falls into this bracket for historical data,

---

[3] *Nuclear Engineering International's World Nuclear Industry Handbook 1992* (Reed Business Publishing Group: London, 1991) publishes reliable data on reactor characteristics; *Nucleonics Week* (McGraw-Hill: New York) publishes monthly generating tables for all commercial reactors, bar those in the former Soviet Union and Eastern Europe.

while fuel discharge projections are published for individual US reactors.[4] Information from the authors' survey of East European fuel-management policy is also used.

2. In the absence of fuel-discharge data there may be information on fuel burn-ups. These may be nominal design equilibrium burn-ups given by reactor (given for reactors in non-nuclear weapon states in IAEA publications), or in the form of historical burn-ups by reactor type over a time period (as is the case for the US data).[5]

3. There may be little fuel and no burn-up information available (as in France). In this case partial evidence from other sources can be used to piece together typical burn-ups or fuelling strategy (for PWRs one-third core reloads and 33 000 MWd/t were assumed for past operation). For France, fuel discharge figures were normalized with reference to published data on total spent fuel discharges to the end of 1989. The gross electrical output and nominal efficiency are known for reactors outside the former Communist world.

4. There may be little reliable data on fuelling strategy, or burn-up, or electrical ouput (as for reactors in the former USSR and the COMECON countries). For these reactors it has been necessary to fall back on a cruder method based on standardized conversion factors which match thermal power output to plutonium production (for example, VVERs are assumed to generate 330 kg of plutonium per GWe(net)-year). Thermal power output is calculated from published design data and assumptions about reactor operating efficiencies.

Clearly, the level of accuracy varies depending on the method used. As a first approximation, errors in the historical estimates (that is, up to 1990) rise from about plus or minus five per cent in method 1 to perhaps plus or minus 10–20 per cent in method 4. The reliability of forecasts is much harder to assess since it depends on the rate of investment in nuclear power as well as the rate at which older reactors are retired. However, the world reactor stock is now quite stable, with an increase in total capacity of less than 10 per cent anticipated over the next decade. Thereafter, capacity will be strongly influenced by the rate of decommissioning of older reactors. Rather than giving 'high' and 'low' estimates, the authors give just one figure. Error bounds for the aggregate figures increase from about plus or minus 10 per cent for the 2000 figures to plus or minus 15 per cent for 2010 figures.

In making forecasts, assumptions have to be made about the evolution of fuel burn-ups in different countries. This will affect both the amount of spent fuel and how much plutonium is produced. Some countries, such as France and Ger-

---

[4] A variety of sources has been used for British figures: Barnham, K. W. J., Hart, D., Nelson, J. and Stevens, R. A., 'Production and destination of British civil plutonium', *Nature*, vol. 317 (19 Sep. 1985), pp. 213–17; *Parliamentary Debates, House of Commons [Hansard]*, parliamentary answers; and the British Department of Energy 'Annual Plutonium Figures', published each autumn (1987–91). US figures are derived from US Energy Information Administration, *World Nuclear Fuel Cycle Requirements 1990*, DOE/EIA-0436(90), Washington, DC, 26 Oct. 1990, pp. 53–70.

[5] For example, the US Energy Information Administration (note 4), pp. 12–14, gives aggregated historical information on US PWR and BWR burn-ups.

Table 5.2. Past discharges of spent fuel and plutonium from power reactors

Fuel discharge is in tonnes of heavy metal; plutonium discharge is in tonnes of total plutonium.

Country	To the end of 1980		1981–90	
	Fuel	Plutonium (Pu$_{tot}$)	Fuel	Plutonium (Pu$_{tot}$)
Argentina	370	1.1	920	3.0
Belgium	200	1.8	1 080	9.7
Brazil	–	–	20	0.2
Bulgaria	150	1.3	600	5.3
Canada	3 480	13.0	11 170	41.6
Czechoslovakia	110	0.4	560	4.9
Finland	40	0.3	720	6.1
France	4 730	11.0	11 220	67.2
FRG	820	6.6	3 740	32.2
GDR	320	2.5	620	5.2
Hungary	–	–	270	2.4
India	190	1.1	550	2.4
Italy	1 280	3.5	820	2.1
Japan	1 470	8.4	6 280	49.0
Korea, South	20	0.2	1 350	8.2
Netherlands	120	1.0	170	1.4
Pakistan	40	0.2	50	0.2
South Africa	–	–	160	1.5
Spain	890	3.2	1 820	10.7
Sweden	370	3.0	2 120	17.7
Switzerland	300	2.5	780	6.7
Taiwan	20	0.2	900	7.4
UK	13 970	27.0	9 010	26.1
USA	6 440	48.7	14 970	127.3
USSR[a]	1 830	7.4	7 570	70.7
Yugoslavia	–	–	110	1.0
Total	**37 160**	**144.4**	**77 580**	**510.2**

[a] Includes Armenia, Lithuania, Russia and Ukraine.

many, look set to pursue higher burn-ups aggressively (leading to lower spent fuel and plutonium discharges), while others will stay with lower burn-ups. Where specific information about burn-up planning is to hand (as for France) it has been used. For those countries with plans to increase fuel burn-up, a set of standard increases is assumed.[6]

[6] Burn-up increases in other countries are assumed to happen only in reactors which operate throughout a given decade (i.e., 2001–10). Germany, Japan, Sweden, Switzerland and Belgium are assumed to raise mean PWR fuel burn-ups to 35 GWd/t in 1991–2000 and to 43 GWd/t in 2001–10. Mean BWR fuel burn-ups are assumed to rise from a mean of 27 GWd/t in 1991–2000 to 35 GWd/t in 2001–10. Mean LWR burn-ups in South Korea, Taiwan and Spain are taken to come into line after 2000. Before that, design burn-ups are assumed.

V. Discharges of spent fuel from civil reactors

At the end of 1990 installed world nuclear capacity was 340 GWe, having nearly tripled over the previous 10 years. By then, the authors estimate that the world's commercial reactors had discharged a total of some 115 000 tonnes of spent fuel, containing a little over 650 tonnes of total plutonium. Spent fuel is currently arising at a rate of around 9300 tonnes each year, and plutonium at about 62 tonnes per annum. We estimate that at the end of 1990 an additional 85 tonnes of plutonium were held in partially irradiated fuel in reactor cores.

Based on a fuelling model for each of the world's power reactors, past spent fuel and plutonium discharges have been calculated. These are set out country-by-country in table 5.2. These figures should be taken as indicative central estimates. In those cases where corroboration of fuel discharge estimates has been possible (for the USA, the UK, France, Germany and Japan) these estimates have been shown to be good, with an error margin of plus or minus 5 per cent.

Plutonium had been generated in power reactors in 26 countries (including the former German Democratic Republic) by the end of 1990. In only 10 of these had more than 10 tonnes been produced. By far the largest power reactor inventory was in the USA, at about 175 tonnes.

Plutonium is currently held in a variety of states and at a variety of sites in each country. Of the material removed from reactor sites, a large proportion has crossed international boundaries to be transported to overseas reprocessing plants. Table 5.2 is therefore not a picture of the actual distribution of plutonium at the end of 1990, although it more nearly represents the present situation in the smaller producers (less than 10 tonnes), than in the larger producers.[7]

Estimating past discharges inevitably involves certain error margins. Making forecasts of spent fuel and plutonium production is even more hazardous, even with a relatively stable number of reactors operating. It is assumed here that all reactors will operate smoothly for their design lifetimes. This projection could be an underestimate. On the one hand, reactor lives may be extended. On the other, it is conceivable that currently unplanned reactors could be operating by 2010, especially in the Pacific Rim countries. Against this, it is possible that some reactors will be retired early, especially in the former Soviet Union and Eastern Europe where there are worries about safety.

With all these provisos, it appears that the rate of increase of fuel discharges has now reached a plateau and is set to fall in the post-2000 period as a result of the combined effect of Magnox reactor decommissioning and higher fuel burn-ups in LWRs. This is taken into account in table 5.3.

The global distribution of power reactor plutonium will change little over the coming 20 years, although the absolute amounts will increase. Spent fuel and plutonium discharges are both set to be about 40 per cent higher during the 1990s than during the 1980s.

---

[7] Switzerland is the main exception.

Table 5.3. Estimated spent fuel and plutonium discharges from power reactors, 1991–2000, 2001–10

Fuel discharge is in tonnes of heavy metal; plutonium discharge is in tonnes of total plutonium.

Country	1991–2000		2001–10	
	Fuel	Plutonium (Pu_{tot})	Fuel	Plutonium (Pu_{tot})
Argentina	1 190	3.9	1750	3.2
Armenia	60	0.2	–	–
Belgium	1 110	10.0	1090	9.5
Brazil	80	0.7	770	7.0
Bulgaria	870	8.1	440	4.4
Canada	16 240	59.9	15 560	60.6
Czech Republic	560	4.9	560	4.9
China	560	5.1	620	5.6
Cuba	100	0.9	140	1.2
Finland	680	5.7	730	5.9
France	13 930	108.6	10 180	109.3
Germany	4 790	44.4	4 180	37.5
Hungary	560	4.9	560	4.9
India	2 590	9.7	2 940	10.2
Italy	140	0.6	–	–
Japan	9 010	83.0	9 760	87.7
Korea, South	3 050	23.1	2 480	19.8
Lithuania	1 300	5.4	1 300	5.4
Mexico	200	1.7	400	3.3
Netherlands	150	1.2	70	0.7
Pakistan	50	0.2	30	0.1
Romania	1 020	3.8	1 700	6.4
Russia	8 430	50.3	8 430	48.9
Slovak Republic	420	3.6	280	2.4
Slovenia	140	1.3	160	1.4
South Africa	480	4.3	480	4.9
Spain	2 380	17.4	1 390	13.3
Sweden	2 340	22.1	2 280	21.7
Switzerland	700	6.8	530	4.5
Taiwan	1 340	11.3	1 330	12.2
Ukraine	3 240	22.4	2 440	20.1
UK	10 620	30.9	5 170	18.8
USA	18 890	181.0	16 540	171.4
Total	**107 220**	**737.4**	**94 290**	**707.2**

Source: Estimates derived by the authors.

Sometime in 1993–94, 35 years after the world's first commercial reactor began operation, 1000 tonnes of plutonium will have been produced in civil power nuclear reactors. (Plutonium discharges will rise from 654 tonnes by the end of 1990 to about 1390 tonnes by the end of 2000.) Over 2000 tonnes of plutonium are expected to be discharged by 2010. Most of this will have been produced in Western Europe, Japan and the USA.

Table 5.4. World discharges of spent fuel from power reactors

Figures are percentage shares.

Region	To 1980	1981–90	1991–2000	2001–2010
Europe[a]	63	43	38	31
North America	27	34	33	34
Pacific rim[b]	4	11	13	15
Former USSR[c]	5	10	12	13
Asia, Africa and Latin America	1	2	4	7
Total	*100*	*100*	*100*	*100*

[a] Includes EC, EFTA and East European countries.

[b] Includes Japan, Taiwan and South Korea.

[c] Includes Armenia, Lithuania, Kazakhstan, Russia and Ukraine.

The geography of spent fuel discharges

To put this array of data into some context, it is useful to look at them in a more aggregate form. We have divided the world into five regions: Europe, North America, the Pacific Rim, the former USSR, and the less-developed countries (China, South Asia, Africa and Latin America). Table 5.4 shows the proportions of decadal spent fuel discharges which the authors estimate to arise in these five regions.

There are two main trends in the distribution of spent fuel arisings: a fall in the proportion of fuel discharges accounted for by European countries from over 60 per cent before 1980 to less than one-third after 2000; and a compensating increase in all other regions. The most significant increases have been in the Pacific rim and the former USSR. The North American proportion remains stable throughout the period.

The reason for the preponderance of European and North American spent fuel discharges can be seen in figure 5.5. Although Magnox reactors represented just 9 per cent of installed nuclear capacity in 1980, they had generated over 55 per cent of the world's spent fuel (about 20 400 t, or some 90 per cent of the European total).[8] Since the majority of this capacity (nearly 80 per cent) was located in the UK, it may be said that, up to about 1980, spent fuel management was a peculiarly British problem. By then British Magnox reactors alone had discharged over one-third of the world's spent fuel (13 700 t out of 37 160 t). It is little wonder that Magnox stations have been described as spent fuel producers which also generate electricity.

Although in the decade 1981–90 the absolute significance of gas-cooled reactors fell, they still accounted for about 44 per cent of European fuel discharges (about 14 300 tonnes out of 32 800 tonnes). Europe therefore continued to be the main spent fuel producer with some 43 per cent of the world total.

---

[8] AGRs had discharged only 250 tonnes of spent fuel by 1980.

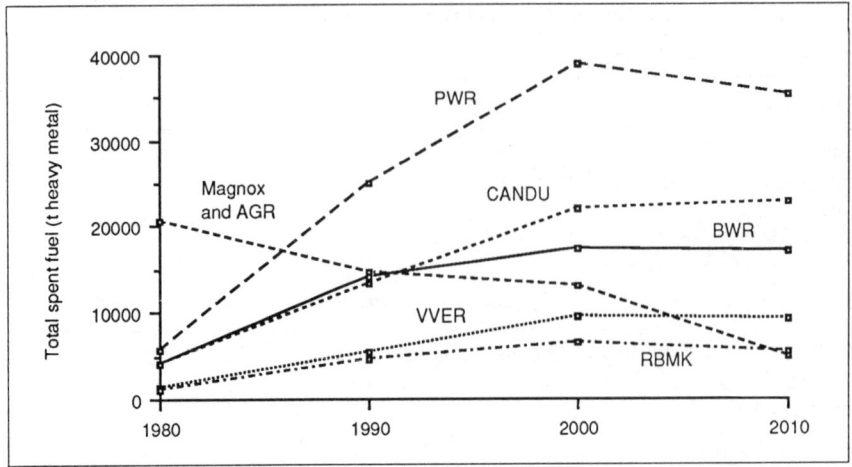

Figure 5.5. Decadal world discharges of spent fuel, aggregated by reactor type, to 1980, 1981–90, 1991–2000, 2001–10

Source: Data derived by the authors.

By then a new, competing trend was becoming clear. Heavy water, natural uranium-fuelled CANDU reactors, the majority of them located in Canada, also generate large quantities of spent fuel. Over the past decade over 40 per cent of North American spent fuel has come from CANDU reactors (11 170 t out of 26 140 t), although they make up only about 10 per cent of the region's nuclear-generating capacity.

These two tendencies—declining Magnox and increasing CANDU capacity—are important to understanding the distribution of the world's spent fuel between 1960 and 2010. Although light-water reactors fuelled with low-enriched uranium have been dominant in terms of capacity, natural uranium-fuelled reactors have most clearly shaped patterns of spent fuel discharges, especially before 1980.

The geography of plutonium arisings

While LWRs produce comparatively small amounts of spent fuel, they are nevertheless responsible for the great majority of plutonium arisings. Natural uranium spent fuel, while bulky, contains less plutonium. The pattern of plutonium production follows the pattern of LWR capacity more closely. The regional distribution of decadal plutonium production figures is shown in table 5.5.

As with spent fuel, the rate of production of plutonium is due to reach a peak during the 1990s, after which there will be a slight decline as a result of shrinking overall capacity and higher fuel burn-ups. The main tendencies evident are

Table 5.5. World plutonium arisings from power reactors

Figures are in tonnes of total plutonium, forecasts rounded to the nearest 5 tonnes.

Region	To 1980	1981–90	1991–2000	2001–2010
Europe[a]	64	199	275	245
North America	62	169	240	230
Pacific Rim[b]	9	65	115	120
Former USSR[c]	7	70	80	75
South Asia, Africa, Latin America	2	7	25	40
Total	**144**	**510**	**735**	**710**

[a] Includes EC, EFTA and East European countries.

[b] Includes Japan, Taiwan and South Korea.

[c] Includes Armenia, Lithuania, Kazakhstan, Russia and Ukraine.

the slight decline in the significance of European plutonium and the rising relative proportion of plutonium produced in the Pacific rim, and in China, South Asia, Africa and Latin America. These latter regions will generate about 4.5 per cent of total plutonium in power reactor spent fuels in the period 1991–2010.

VI. Conclusion

When it is discharged, spent fuel must be stored under safe and secure conditions. Initially this is invariably at the reactor site. There are then two options for the operator: to extend storage capacity on-site; or to send it to a reprocessing plant. In the next chapter we assess historic and forecast rates of reprocessing. A third option, to send the fuel to a dedicated 'away from reactor' (AFR) store, is also available in Sweden and is planned in Germany.[9]

While in store, spent fuel remains a proliferation and safety risk. Accidents or diversions of material are possible up to the point when the fuel is finally disposed of underground, and may even continue after that. Continued institutional supervision of spent-fuel wastes above ground is likely to last at least 20 years, and could go on for much longer. The main reason for waiting is that the decay in heat production will make geological disposal less hazardous and costly. Throughout the long periods in which spent fuels may be held in store, they will have to be subjected to strict safeguarding and physical protection.

Spent fuel will continue to accumulate at a rate of 9000–10 000 tonnes per year for the next two decades. Each year a further 60–70 tonnes of plutonium will be added to the world total. The overwhelming majority will continue to arise in Europe, North America and the Pacific Rim, although the amounts now being produced in the transitional and less developed world are reaching the hundreds of tonnes. Maintaining control over this material presents a major challenge for safeguards regimes.

[9] *The Nuclear Fuel Cycle Information System: A Directory of Nuclear Fuel Cycle Facilities,* IAEA, STI/PUB/794, Vienna, 1988, pp. 33–35.

6. Reprocessing programmes and plutonium arisings

By the end of 1990 about 650 tonnes of plutonium had been discharged from the world's power reactors. Most of this material was still fixed in spent fuel, but a substantial and growing amount had been separated. In this chapter estimates are made of how much plutonium has been separated at reprocessing plants to date from power-reactor fuel, and how much is due to be separated over the next two decades.

I. Reprocessing in the nuclear fuel cycle

Reprocessing is the chemical separation of plutonium (0.2 to 1 per cent by weight of the irradiated fuel) and uranium (over 95 per cent by weight of the fuel) from the fission products and transuranic wastes contained in spent nuclear fuel. Three main reasons have historically been put forward for spent fuel reprocessing.

First, it makes available fissile materials which can be recycled as fuel in thermal and fast reactors.

Second, it is a safer way of dealing with the spent fuel. This is especially so with metallic uranium fuel used in Magnox reactors, whose cladding corrodes rapidly when stored under water. Splitting the fuel from its cladding and dealing with the waste products separately may, in this case, be safer than trying to deal with the untreated fuel rods.

Third, fuel is taken off the reactor site where it would otherwise accumulate. In practice this has become the major driving force behind most utilities' reprocessing policies. Reactor operators have been keen to be rid of the problem of managing spent fuel, so long as the cost of reprocessing does not greatly affect the overall cost or public acceptability of nuclear electricity. Typically, a utility reactor operator will have a contract with a fuel services company (such as Cogema in France or British Nuclear Fuels (BNFL) in the UK) for a specified amount of spent fuel to be reprocessed. All the components of the fuel remain the property of the utility, and in due course they are returned.

If plutonium and uranium are to be reused, they must be in a chemically pure form, free of fission products and other materials. To do this, fuel cladding is first removed from the fuel, either mechanically (as with metallic fuel) or chemically (as with oxide fuel) and the fuel is dissolved in nitric acid (see figure 5.1). It is at this stage that the first accurate measurement can be made of the amount of plutonium contained in the fuel. Radiations emitted by spent fuel can be measured by non-destructive assay methods, but these do not yield the same accuracy as weighing and analytical techniques which can be applied to solutions in the process areas of a reprocessing plant. The plutonium content in spent fuel cannot be measured directly, but is estimated by utilities using their knowledge of reactor operating histories.

The dissolved fuel is put through a series of *solvent extraction* phases in which the fission products and transuranics are first separated out, followed by the uranium and plutonium (the reprocessing products). Uranium is usually made available to the customer as a nitrate, whereas plutonium is normally returned as an oxide.

There are inevitably losses of material during decladding and chemical processing. In the overall material balance of the Eurochemic plant in Belgium which operated from 1966 until 1974, some 2.3 per cent of the uranium and 4.8 per cent of the plutonium were lost to waste streams.[1] At Sellafield it has been estimated that between 2 and 3.5 tonnes of plutonium are contained in a variety of plutonium contaminated solid wastes which have accumulated at the site over the past 40 years.[2] A 1986 government report showed that about 1.8 tonnes of plutonium were held in stripped fuel cladding and other plutonium contaminated materials (PCM) produced in magnox reprocessing.[3] This is equivalent to losses of 4–5 per cent in magnox reprocessing.[4]

Losses have generally been reduced with improved technology and tighter regulatory control on waste management. Today, overall plutonium losses at oxide fuel reprocessing plants probably lie between 0.5 and 1.5 per cent of the input total. In making estimates here we assume losses of 1–2 per cent for oxide fuel reprocessing (depending on the plant) and 2–5 per cent for magnox fuel reprocessing.[5]

II. The evolution of fuel-cycle strategies

The 1960s

Irradiated fuel was first discharged from commercial power reactors in the early to mid-1960s in the USA (Dresden 1 first discharged fuel in 1961), the UK, the Soviet Union and France. In these early days, nuclear fuel remained the property of state authorities in charge of nuclear power development—the US Atomic Energy Commission (AEC), the United Kingdom Atomic Energy Authority (UKAEA), the Soviet Ministry of Atomic Power and Industry

---

[1] Detilleux, E., 'Operation of the plant and the period after shutdown', eds W. Drent and E. Delande, *Proceedings of the Seminar on Eurochemic Experience,* 9–11 June 1983, ETR-318, Eurochemic, Mol, Belgium, Apr. 1984, p. 59.

[2] Hinkley Point 'C' Inquiry, Transcript of evidence day 98, UK, 1989, p. 17.

[3] HMSO, *Radioactive Waste Mangement Advisory Committee Seventh Annual Report,* London, Sep. 1986, figure 3.

[4] These large losses are chiefly the result of the mechanical decladding of magnox fuel. Selective chemical decladding used for oxide fuels tends to reduce plutonium losses to ≤1 per cent.

[5] For a discussion of magnox reprocessing losses see Barnham, K. W. J., 'Calculating the plutonium in spent fuel elements', ed. F. Barnaby, *Plutonium and Security* (Macmillan: London, 1991), pp. 110–32.

(MAPI)[6] and the Commissariat à l'Énergie Atomique (CEA). In all but the Soviet case, power-reactor fuel was sent for reprocessing from the earliest date.

In the UK and France reprocessing lines originally devoted to plutonium production for weapons were used to process fuel from commercial Magnox reactors. Some plutonium contained in this fuel was appropriated for weapon purposes (see chapter 3). It was not until the early 1970s that utilities in these countries began to assume ownership of their fuel, and that a clearer distinction was drawn between what was civil and what was military.

The fuel arisings from commercial reactors in non-nuclear weapon states have been kept out of military programmes, with one possible exception. Spain's Vandellos 1 Magnox reactor, which operated from 1972 to 1990, was supplied by France which also provided all the fuel on a 'take-back' basis.[7] Plutonium separated from early fuel discharges from this reactor may have been used in the French weapon programme.

A similar scheme of supplying and taking back fuel was operated until the late 1980s by the Soviet Union for the reactors it supplied to Finland and to countries in Eastern Europe. VVER fuel from domestic and foreign reactors has been reprocessed since 1978. The plutonium separated was to have been used to fuel the former USSR's fast reactor programme, but this has now been curtailed. Substantial amounts of reprocessed uranium (RepU) have already been recycled, as in the UK.[8] RepU was used to fabricate new fuel for the Soviet Union's own RBMK reactor programme (a reactor type never exported).[9]

The 1970s

Between 1970 and 1980 world installed nuclear capacity rose from around 10 GWe to 120 GWe. During this period, most reactors were built with limited on-site fuel storage capacity. A typical Westinghouse PWR, for instance, was supplied with storage capacity sufficient for about four years'-worth of fuel discharges, while Soviet VVERs were built with five to six years' storage capacity. Provision therefore had to be made to remove fuel, usually to a reprocessing site. Up to 1975 it was assumed that most spent reactor fuel would be sent for reprocessing. The exceptions were CANDU and RBMK reactors, for which long-term fuel storage was seen as the preferred option. India alone has reprocessed CANDU fuel.

For this reason, there was an attempt by countries with ambitious nuclear policies to match spent fuel discharges from light-water and gas-cooled reactors

---

[6] Sometimes called the Ministry of Nuclear Power (MNP).

[7] Magnox reactors sold by the UK and Italy were fuelled under a less restrictive arrangement. Spent fuel was sent to Windscale/Sellafield for reprocessing, but the plutonium was made available for return.

[8] Over 15 000 t of depleted magnox RepU has been used as feed for the Capenhurst diffusion enrichment plant. Forsey, D. C. and Gresley, J. A. B., 'An outline of the requirements for successful uranium recycle', Paper given to RECOD 87 Conference, Paris, Aug. 1987.

[9] Separated plutonium and waste were retained and stockpiled at the reprocessing site. In the late 1980s the Soviet Union tried to put its fuel contracts with foreign customers on a commercial footing. This meant setting market prices on reprocessing services and insisting on the return of reprocessing products and wastes for the first time.

with reprocessing capacities. Often this led to optimistic plans being drawn up for domestic reprocessing plants.[10] Such plans usually included a commitment to large-scale plutonium recycling in fast reactors. However, this vision of a plutonium economy did not hold for long. From the mid-1970s onwards, reduced expectations for the growth of nuclear electricity generation, the abundance of uranium and delays in commercializing fast reactors undermined the rationale for rapid and complete reprocessing.

Doubts were also raised over the commercial viability of reprocessing by the failure of a privatized US reprocessing industry to get off the ground in the mid-1970s.[11] Reprocessing of commercial reactor fuel was effectively abandoned in the USA after 1976, partly as a result of the Carter Administration's policy to achieve a world-wide ban on reprocessing in its campaign against nuclear proliferation. Spent fuel has subsequently been stored at reactor sites. Most reactor operators have adapted to the change in policy by extending fuel storage capacities, either by stacking fuel more tightly into existing ponds, or by making available new capacity such as fuel cask and vault systems.[12]

The 1980s and 1990s

In contrast, Japan and European countries with LWR programmes have, in most cases, persisted with reprocessing policies. A mixture of strategic, political and legal factors have been responsible. As a result, a majority of Japanese and West European spent fuel (about 80 per cent of discharged fuel containing about 50 per cent of discharged plutonium) discharged by the year 2000 is due to be reprocessed at four centres—Sellafield in the United Kingdom, La Hague and Marcoule in France, and Tokai-mura in Japan. The first two of these centres are much the largest, and they have become hubs in a global fuel service industry.

A strong connection has developed between European and Japanese fuel-cycle policies. Almost two-fifths of LWR reprocessing capacity available to non-British and non-French utilities over the coming decade will be taken up with Japanese fuel.[13] This European–Japanese fuel management regime has existed since the mid-1960s when magnox fuel from the Tokai reactor was first shipped to Windscale, but it is now showing signs of strain. A growing proportion of European and Japanese fuel will be left in store at reactors or in central stores in future years, and plutonium appears likely to be viewed less as a fuel than as a disposal problem.

---

[10] Japan, FR Germany, Belgium and Sweden all put forward plans for reprocessing facilities during this period, while France, the UK and the USA planned to expand their capacity.

[11] Three US reprocessing facilities, at West Valley, Barnwell and Morris, closed down or failed to come into operation during the mid-1970s. See Rochlin, G., *Plutonium, Power and Politics* (University of California Press: Berkeley, Calif., 1979), pp. 104–105.

[12] *Nucleonics Week* and *Nuclear Fuel*, 'Spent fuel storage options', Special Report, 27 Oct. 1988.

[13] See Berkhout, F., Suzuki, T. and Walker, W., 'The approaching plutonium surplus: a Japanese–European predicament', *International Affairs,* vol. 66, no. 3 (July 1990), pp. 523–45.

Table 6.1. National spent-fuel management policies, 1960–2000 and beyond

Country	1960s/1970s	1980s	1990s	Post-2000
Argentina	S	S	S + R (D)?	S + R (D)?
Armenia	. .	. .	?	?
Belgium	R (D)	R (F + D)	R (F) + S	S?
Brazil	S	S	S	S
Bulgaria	. .	TB	S?	S?
Canada	S	S	S	S
China	. .	. .	S?	S?
Czech Republic	. .	. .	R(F) + S?	S?
Czechoslovakia	. .	TB	R (F) + S?	S?
Finland	. .	TB + S	TB + S	S?
France	R (D)	R (D)	R (D) + S	R (D) + S?
FRG	R (F + D)	R (F + D)	R (F) + S	R (F)? + S
GDR	TB	TB	. .	. .
Hungary	. .	TB	R (F)	R (F)? + S
India	R (D)	R (D)	R (D) + S?	R (D) + S?
Italy	R (F)	R (F)	R (F)	. .
Japan	R (F)	R (F + D)	R (F + D) + S	R(F?+D) + S?
Kazakhstan	. .	. .	?	?
Korea, South	. .	S	S + R (F)?	S + R (F)?
Lithuania	. .	. .	S?/R (F)?	S?/R (F)?
Mexico	. .	. .	S	S
Netherlands	R (F)	R (F)	R (F) + S	S
Pakistan	R (D)?	R (D)?	R (D)?	?
Romania	. .	. .	S	S
Russia/USSR	R (D) + S	R (D) + S	R (D) + S?	R (D)? + S
Slovak Republic	. .	. .	R(F) + S?	S?
Slovenia	. .	S	S?	S?
South Africa	. .	S	S	S
Spain	. .	R (F) + TB	S	S
Sweden	R (F)	S	S	S
Switzerland	R (F)	R (F)	S	S
Taiwan	. .	S	S	S
Ukraine	. .	. .	S?/R (F)?	S?/R (F)?
UK	R (D)	R (D)	R (D) + S	R (D) + S
USA	R (D) + S	S	S	S

Note: S = storage; R (D) = reprocessing (domestic); R (F) = reprocessing (foreign); TB = fuel returned to supplier under 'take-back' arrangement.

Fuel-management policies in the former Soviet Union have become even more troubled. Hitherto, all spent fuel has been sent to reprocessing and storage sites at Chelyabinsk or Krasnoyarsk, both of which are in Russia. It remains to be seen whether spent fuel from VVERs in Eastern Europe, the Ukraine, Lithuania and Russia itself will continue to be sent to these sites, or will instead be stored at reactor sites, and whether reprocessing at Chelyabinsk will continue.

A summary of fuel-management strategy in countries with major nuclear programmes is given in table 6.1. It shows that in several countries there is considerable uncertainty about the policies that will be followed in future.

III. A sketch of methods

Making estimates of the amount of plutonium separated at reprocessing plants is even more problematic than calculating plutonium production in reactors. Complete information on fuel throughputs, plutonium concentrations and material losses at reprocessing facilities is usually not in the public domain. Instead, a number of assumptions have to be made, often based on partial or aggregated information. Three different approaches have been used.

1. In the best cases, annual throughputs of fuel and their mean burn-ups are known. For no plant is a full set of data available, but for some—UP1 and UP2 in France, Tokai-mura in Japan, and B205 at Sellafield in the UK (since 1981)—a good approximation can be made. In some cases, reprocessors have published cumulative totals of plutonium separated.

2. Information on totals of spent fuel reprocessed at a particular plant up to a certain date may be used, together with inferences about burn-ups which can be corroborated with estimates made for discharged fuel from individual reactors (see chapter 5).

3. There may be information about the capacity of the plant and the reactors which it is servicing (as with B205 before 1981, and Soviet and Indian reprocessing plants). In this case, using information about spent-fuel discharges from reactors or about the sizes of contracts, rates of plutonium separation can be inferred.

The basic data used in this section are set out in appendix C. We have tried to state all final quantities in terms of total plutonium. When converting from fissile to total plutonium quantities, the assumption is made that the material was 70 or 80 per cent fissile if produced in LWR or Magnox reactors respectively. This does not take into account the subsequent decay of plutonium-241 (half-life: 14.4 years).

IV. Overview of power-reactor fuel reprocessing

Power-reactor fuel has been reprocessed at numerous facilities around the world, several of them small plants. In this chapter only the most important in terms of plutonium separation are considered, namely those which have separated at least 1 tonne of plutonium. These facilities are described in table 6.2 which presents design annual throughputs for reprocessing plants.

Within the industry, daily throughputs are more usually cited for these plants, together with an estimate of how many days the plant is expected to operate in

Table 6.2. Commercial reprocessing plants around the world

Capacity is in tonnes of heavy metal per year.

Country	Location	Owner/ Operator[a]	Facility[b]	Fuel[c]	Design capacity	Year of operation
UK	Windscale/ Sellafield	BNFL	B205	metal	1500	1964–2010?
			B204/205	oxide	300	1969–73
			THORP	oxide	700	1993–
	Thurso	UKAEA	DNPDE	oxide (MTR)	< 1	1959–1996?
				oxide (FBR)	7	1958–95?
France	Marcoule	Cogema	UP1	metal	400	1958–2000?
	La Hague	Cogema	UP2	metal } oxide }	400	{ 1966–87 { 1976–90
			UP3	oxide	800	1990–
			UP2–800	oxide	800	1992–
	Marcoule	CEA	APM	oxide (FBR)	6	1988–
USA	West Valley	NFS	West Valley	oxide	300	1966–72
Russia[d]	Chelyabinsk-40	MAPI	Mayak	oxide	600	1978–
Japan	Tokai-mura	PNC	Tokai	oxide	90	1981–
	Rokkasho-mura	JNFS	Rokkasho	oxide	800	2002–
India	Tarapur	DAE	PREFRE	oxide	100	1982–
	Kalpakkam	DAE		oxide	100–200	1993/94–
FRG	Karlsruhe	KfK/DWK	WAK	oxide	35	1971–90
Belgium	Mol	Eurochemic		oxide	30	1966–74

[a] BNFL = British Nuclear Fuels plc; UKAEA = UK Atomic Energy Authority; Cogema = Compagnie Générale des Matières Nucléaires; CEA = Commissariat à l'Énergie Atomique; NFS = Nuclear Fuel Services Company; MAPI = Ministry of Atomic Power and Industry; PNC = Power Reactor and Nuclear Fuel Development Corporation; JNFS = Japan Nuclear Fuel Service Company; DAE = Department of Atomic Energy; KfK = Kernforschungszentrum Karlsruhe; DWK = Deutsche Gesellschaft für Wiederaufarbeitung von Kernbrennstoffe.

[b] THORP = Thermal Oxide Reprocessing Plant; DNPDE = Dounreay Nuclear Power Development Establishment; APM = Atelier Pilote Marcoule; PREFRE = Power Reactor Fuel Reprocessing; WAK = Wiederaufarbeitungsanlage Karlsruhe.

[c] LWR fuels unless stated otherwise. MTR = materials test reactor; FBR = fast breeder reactor.

[d] According to reports, construction was begun on another plant at Krasnoyarsk (Dodonovo-27), but was halted by environmental opposition having been 30% completed. It is assumed here that the plant will not be completed. See Cochran, T. B. and Norris, R. S., 'A first look at the Soviet bomb complex', *Bulletin of the Atomic Scientists*, May 1991, pp. 25–31.

a given year. Down-times between reprocessing campaigns for maintenance and retrofitting are relatively lengthy (a typical plant may be operable for 150–200 days each year) and are the main factor determining annual throughput. A reduction in down-times allows annual throughputs to be raised.

V. Commercial reprocessing programmes

In many cases, reprocessing plants have not achieved design fuel throughputs. They have frequently incurred technical problems which have either restricted production or forced lengthy closures while changes were made to the plant, or both. The most consistent load factors have been achieved in plants processing low burn-up fuel from Magnox reactors. Historically, plutonium separation rates have fallen below what was planned. In this section a detailed analysis is given of the operating histories of commercial-scale reprocessing plants. Summary tables of the information presented here are given in section VI.

The United Kingdom

Magnox fuel

Separation of weapon plutonium began at Windscale/Sellafield in 1951 at the Butex B204 plant. Fuel from two plutonium production reactors (the Windscale Piles) was reprocessed until they were shut down in 1957. From 1956 on, fuel from the eight dual-purpose plutonium- and electricity-producing Magnox reactors at Calder Hall and Chapelcross was also reprocessed at B204. Maximum throughput at the plant was about 750 tonnes of low burn-up fuel per year until it was shut down in 1964.

Magnox power-reactor fuel has been reprocessed at the follow-on plant (B205) since 1964—one year after fuel was first discharged from the Magnox station at Berkeley. Processing of fuel from Calder Hall and Chapelcross was also transferred to B205, where it was 'co-processed' with power-reactor fuel (that is, in mixed reprocessing campaigns) until 1986.

B205 has always been a critical element of the Magnox programme. Not only were fuel storage capacities at reactors limited, but the onset of cladding corrosion made the fuel more hazardous to handle and transport if stored in ponds for over 12 months. The smooth transfer of fuel to Windscale/Sellafield was therefore vital to continued reactor operation. When production at B205 was halted in the early 1970s, Magnox reactor operation was put in jeopardy.[14]

Two factors have significantly reduced this pressure in the past decade. First, Magnox reactors have operated at higher burn-ups, and have thus discharged less fuel. Second, the control of pond water chemistry has improved, so reducing rates of cladding corrosion. The typical lag between fuel discharge and reprocessing today is over two years.

Part of the capacity at B205 was taken up by fuel from the two exported Magnox reactors in Italy and Japan. When operating normally, these reactors

---

[14] Two problems developed at Windscale/Sellafield. First, storage capacity for magnox spent fuel was becoming exhausted. Second, and more important, liquid radioactive discharges from Windscale/Sellafield and from reactor sites were reaching their authorized limits because increasingly contaminated pond water was being washed out to sea (or into a lake in the case of Trawsfynydd).

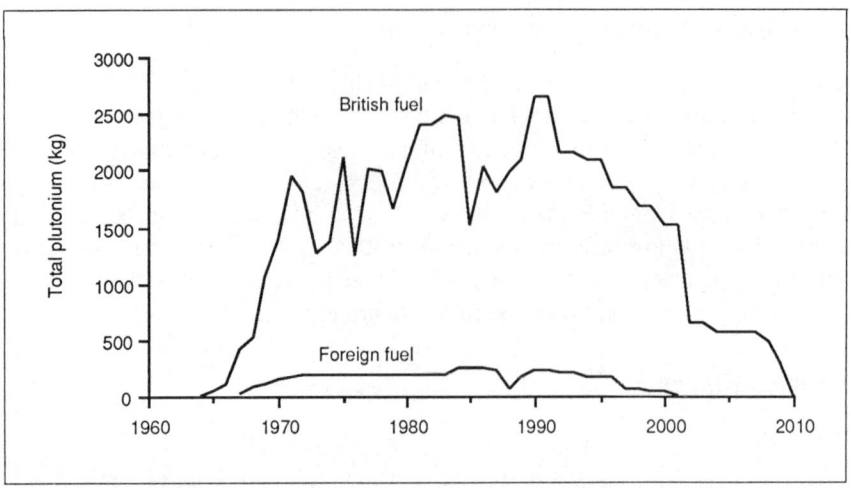

Figure 6.1. Past and projected quantities of magnox plutonium separated at the British B205 reprocessing plant, British and foreign fuel, 1964–2010

Sources: Parliamentary Debates, House of Commons [Hansard], Official Report, 1 Apr. 1982, cols 168–69; 27 July 1983, col. 440; 25 Jan. 1985, cols 545–46; 23 July 1985, col. 413; 21 July 1986, col. 10; Department of Energy, Press Releases, 'Annual Plutonium Figures', 16 Dec. 1987; 13 Oct. 1988; 5 Dec. 1989; 18 Oct. 1990; 17 Oct. 1992; and Barnham, K. W. J., Hart, D., Nelson, J. and Stevens, R. A., 'Production and destination of British civil plutonium', *Nature*, vol. 317 (19 Sep. 1985), pp. 213–17.

typically discharged about 100 tonnes of fuel annually. This fuel was routinely shopped to Sellafield after about a year's cooling at the reactor. The Latina reactor was decommissioned in 1987, and the Tokai unit is due to follow suit in 1995. Reprocessing of foreign magnox fuel is therefore set to end in about 2000.

Basic operating information on fuel throughput and plutonium separation for the B205 plant has been published since 1982. For the period 1964–81 it has been necessary to make certain assumptions about the rate of reprocessing. The work of Barnham and colleagues has provided us with fairly good fuel discharge data have been generated for the Magnox programme.[15] These can be brought together with the fuel discharge estimates derived for the Tokai and Latina reactors. An axiom in deriving plutonium separation estimates is that all magnox fuel is reprocessed soon after discharge from the reactor. By assuming a universal one-year lag between fuel discharge and reprocessing before 1981, and by adopting the mean fuel burn-ups calculated for these reactors (ranging from 3000 MWd/t before 1980 to 4900 MWd/t at present, reliable estimates can be derived for plutonium separation. Inevitably the accuracy of estimates for individual years may vary in the pre-1981 period (for instance no account

---

[15] Barnham, K. W. J., Hart, D., Nelson, J. and Stevens, R. A., 'Production and destination of British civil plutonium', *Nature*, vol. 317 (19 Sep. 1985), pp. 213–17.

Table 6.3. Cumulative plutonium separation at the British B205 plant, ends of 1970, 1980 and 1990

Figures are in kilograms.

	End of 1970	End of 1980	End of 1990
British	3 500	20 950	42 450
Since 1971		15 500	37 000[a]
Foreign	350	2 200	4 200
Total	**3 850**	**23 150**	**46 650**

[a] This figure compares well with the Department of Energy's published figures (see text).

has been taken of the temporary halt in production at B205 during 1973), but we have good confidence in the cumulative figures. Figure 6.1 gives estimates of plutonium annually separated from magnox fuel at Windscale/Sellafield asuning steady reprocessing of fuel.

In the 20 years since 1970, plutonium separation has fluctuated between 1.2 and 2.6 tonnes per year, with a mean of 2.14 tonnes. British material made up just over 90 per cent of this total. Plutonium production is now expected to fall away (from a mean of 2.35 tonnes/year in the 1980s to about 1.9 tonnes/year in the 1990s) as a result of reactor closures. Magnox reprocessing is expected to end in about 2010 if current plans for life extension of certain reactors are realized. By then a total of about 72 tonnes of plutonium may have been separated from Magnox power-reactor fuel at B205, 5.4 tonnes of which were discharged at the Japanese and Italian reactors.

Aggregate totals for plutonium separation based on these figures are set out in table 6.3. All figures are given for the end of the financial year (31 March), as is the British convention. The more important of the British figures is the cumulative total separated since 1971/72 when title to nuclear fuel, including plutonium, was moved from the UKAEA to the utilities. Published annual accounts use 1971/72 as the base year. Nearly 5.5 tonnes of British power reactor plutonium were separated from power-reactor fuel before then and it is presumed that most of this material was bartered with the USA (see chapter 3).

The figure derived (37 tonnes) for plutonium separation up to the end of 1990 compares well with the British Department of Energy's published figures. The Department of Energy, *News Release 195,* 18 October 1990 states that at 31 March 1990, 31 tonnes of separated Magnox plutonium were in store at Sellafield, that 0.5 tonne was in process and that 5.5 tonnes had been sold or leased to the UKAEA for fast reactor research (albeit since 1969, not 1971 which is used as the datum here). This gives a total of 37 tonnes.

On 31 March 1991, 33 tonnes of British magnox plutonium and 1 tonne of foreign-owned plutonium were in store at Sellafield. About 5 tonnes are in use by the UKAEA for fast reactor R&D. At current rates of production there will be a stock of 50 tonnes of plutonium separated from British reactor fuel at

Table 6.4. Existing contracts for reprocessing at THORP, UK, 1992

Figures are in tonnes of heavy metal.

Country	Fuel contracted	
	1992–2002	Post–2002
UK		
Nuclear Electric	1 500	1 520
Scottish Nuclear	350	330
AEA Technology	(150)[a]	–
FRG	968	1 500[b]
Italy	157	–
Japan	2 680	–
Netherlands	55	–
Spain	169	–
Sweden	(139)[c]	–
Switzerland	461	–
Total	**6 340**	**3 350**

[a] Letter of Intent signed.

[b] These contracts, signed in 1990, contain a *force majeure* clause allowing German utilities to pull out of them in certain circumstances, such as a change of government policy. Previous contracts had been binding on the utilities.

[c] Contracts signed in 1979. Subsequently there was an abandonment of reprocessing by Swedish utilities, and no fuel has been sent to Sellafield under the contract.

Sources: Albright, D., *World Inventories of Plutonium*, PU/CEES Report no. 195, Center for Energy and Environmental Studies, Princeton University, Princeton, N.J., June 1987, pp. 57–58; Marshall, P., 'BNFL to charge about 500 pounds/kg for reprocessing in German contracts', *Nucleonics Week*, 3 Sep. 1990; Buxton, J., 'Scottish Nuclear to cut fuel reprocessing needs', *Financial Times*, 11 Feb. 1992.

Sellafield by 2000, and this will have risen to about 56 tonnes by the end of magnox fuel reprocessing in 2010 (assuming 5 tonnes remain in fast reactor fuel). If the fast-reactor programme is wound up as planned by the end of 1994, and the material leased by the UKAEA is returned, the surplus should be 61 tonnes.

Oxide fuel

Oxide fuel reprocessing at Windscale/Sellafield began in 1969. A new head-end plant (HEP) at which fuel was chopped up and dissolved was sited at the idle B204 plant. From there the fuel solution was passed through one solvent extraction cycle in the old Butex plant, before being fed on a campaign basis into the B205 plant.[16] In all, 73 tonnes of oxide fuel were processed in this way before an accident caused the plant to be shut down in 1973. Most of this

[16] Hudson, P., 'Developing technology to reprocess oxide fuel', *British Reprocessing, Nuclear Engineering International* Special Publication, Oct. 1990, pp. 17–20.

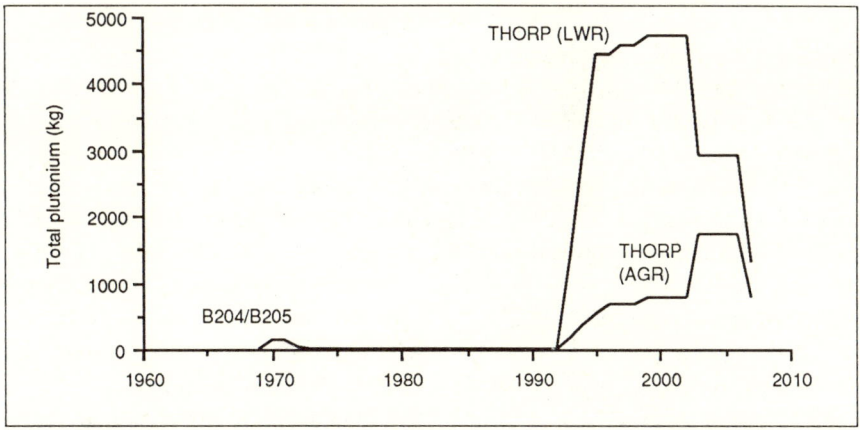

Figure 6.2. Past and projected plutonium separation from oxide fuel at the British Sellafield B204 and B205 plants, and signed and agreed contracts for THORP capacity, 1969–2007

Note: Reprocessing contracts already signed for the post-2002 period are included here. The assumption is that THORP will continue to operate at full capacity until these contracts are fulfilled.

Sources: Hudson, P., 'Developing technology to reprocess oxide fuel', *British Reprocessing, Nuclear Engineering International* Special Publication, Oct. 1990, pp. 17–20; and table 6.4.

fuel was from European research and materials testing reactors. Some 360 kg of plutonium were extracted.

Large-scale thermal reactor oxide fuel reprocessing is not due to begin until late 1992 when the new Thermal Oxide Reprocessing Plant (THORP) begins operating at Sellafield. About two-thirds of THORP capacity in the first 10 years of operation will be dedicated to handling foreign spent fuel. The contractual position as it stood at the end of January 1992 is set out in table 6.4.

Past and projected rates of plutonium separation from oxide fuel at Sellafield are shown in figure 6.2. Assuming that THORP is commissioned as planned, there will be a rapid escalation of oxide plutonium production from 1992, quickly passing the amount separated annually from magnox fuel. By the late 1990s THORP is expected to be producing about 5.5 tonnes per year, which is over twice the rate ever achieved with magnox reprocessing. By 2002, THORP is expected to have separated about 47 tonnes of plutonium, 41 tonnes of which will have come from foreign LWR fuels. All of this material is due to be returned to the country of origin.

Fast reactor fuel

Oxide FBR fuel has been reprocessed at Dounreay since July 1958. By the end of 1990 a total of about 24 tonnes of fuel had been processed at the facility. Of this 10 tonnes were spent HEU fuel containing no significant quantities of plutonium discharged by the Dounreay Demonstration Fast Reactor (DFR) before it was shut down in 1977. A further 14.35 tonnes of plutonium/uranium mixed oxide fuel from the Prototype Fast Reactor (PFR) has been processed since 1980. In total, 3.08 tonnes of plutonium had been separated from this fuel by the end of 1990.[17] This makes up about two-thirds of the amount (5 tonnes) in the PFR fuel cycle.

The PFR was brought into operation in 1974. Since then, as with all research reactors, it has operated unevenly, achieving a cumulative load factor of 23.5 per cent.[18] If fuel burn-ups had reached design levels (61 000 MWd/t), only about 13.5 tonnes of fuel would have been discharged (assuming about half the core is replaced at each refuelling).[19] Since more fuel has been discharged, fuel burn-ups must have been below design levels. In these circumstances estimates of future fuel discharges and reloads are very difficult. If we assume a mean cooling time for PFR fuel of three years, then it can be estimated that some 17.9 tonnes of fuel had been discharged by the end of 1990 containing 3.8 tonnes of plutonium. Adding to this the amount of plutonium in the reactor core—880 kg—it can be estimated that a total of 4.65 tonnes of plutonium has so far been inserted into the PFR core. This suggests that little of the recovered plutonium has so far been recycled in the PFR.

The future of PFR and the Dounreay reprocessing plant is unclear (see section IV, chapter 7). If PFR matches its past performance in the period to 1994 and is then decommissioned, and if all its fuel is reprocessed, some 5.4 tonnes of plutonium could have been separated by 1997/98.[20] Future reprocessing of MTR fuel at Dounreay will not yield significant quantities of plutonium.

France

No clear distinction is made in France between military and civil nuclear materials.[21] Indeed the French have been open about their use in weapons of

---

[17] Barrett, T. R., 'Specialist reprocessing needs', paper given to the Management of Spent Nuclear Fuel Conference, IBC Technical Services, London, 29–30 Apr. 1991. Total throughput at 31 Dec. 1986 was put at 20.3 t of fuel (10.06 t from the DFR and 10.3 t from the PFR). See Mégy, J., Allardice, R. H., Ebert, K., Morelle, J.-M. and Venditi, P., 'The fast breeder reactor fuel cycle in Europe—present status and prospects', *Nuclear Technology*, vol. 88 (Dec. 1989), pp. 283–89.

[18] *Nuclear Engineering International, World Nuclear Industry Handbook 1992* (Reed International: London, 1991), p. 21. For a list of more technical information about DFR and PFR, see table 7.1.

[19] The PFR core contains 4.1 t of fuel. Discharging half that amount for 16 years would yield 32.8 t of fuel. At a capacity factor of 23.5% just 7.7 t would have been discharged.

[20] This assumes that a further 3.3 t of fuel containing 21.5% plutonium will be loaded between 1990 and 1993.

[21] The main references used in this section besides those noted later are: Commissariat à l'Energie Atomique (CEA), 'Le retraitement des combustibles irradiés', *Industrie Nucleaire Française*, Paris, 1982 pp. 154–64; Couture, J., 'Status of the French reprocessing industry', paper given to American Nuclear Society Conference, *Fuel Processing and Waste Management*, 26–29 Aug. 1984, Jackson, Wyo.; Delange, M., 'Operating experience with reprocessing plants', *Atomwirtschaft*, Jan. 1985, pp. 24–28; Delange, M., 'LWR spent fuel reprocessing at La Hague: ten years on', *RECOD 87* conference, Paris, 1987, pp. 187–93; 'Reprocessing and waste management, country: France, pt 1', *NUKEM Market Report*, no 3 (1988), pp. 15–18; Lewiner, C. and Gloaguen, A., 'The French reprocessing programme', *Atomwirtschaft*, May 1988, pp. 227–29; CEA, *Cycle du combustible nucléaire: retraitement*, Paris, Mar. 1989; 'Reprocessing and waste management: review 1989', *NUKEM Market Report*, no. 2 (1990), pp. 14–23; EdF, 'Retraitement recyclage', Paper by Service des Combustibles, Paris, 6 Mar. 1990.

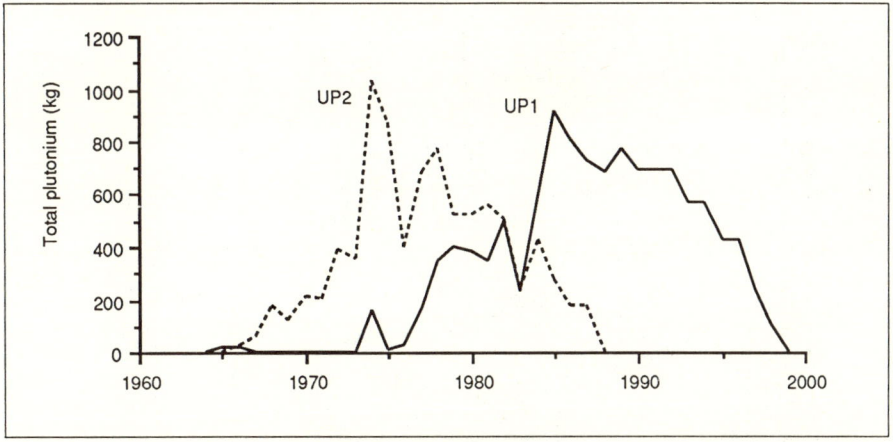

Figure 6.3. Past and projected plutonium separation from Magnox power-reactor fuel at the French Marcoule UP1 and La Hague UP2 reprocessing plants, 1965–2007.

Sources: Syndicat CFDT de l'Énergie Atomique, *Le dossier électronucléaire* (Éditions du Seuil: Paris, 1981), pp. 186–91; Hirsch, H. and Schneider, M., 'Wiederaufarbeitung in Europa: Wackersdorf ist tot es—lebe La Hague?', *Rest-Risiko*, no. 6 (Greenpeace: Hamburg, Apr. 1990).

plutonium produced in power reactors operated by Électricité de France (EdF) (section VI, chapter 3). This section traces the reprocessing of fuel from all power reactors, whether or not it has ended up in military or civil use. Estimates of how much of this material may have been put to weapon use are relatively low and are given in chapter 3.

Magnox fuel

Plutonium separation began in France in November 1949 when French chemists at the Le Bouget centre succeeded in isolating 15 milligrams from a fuel rod irradiated in the Zoé heavy water reactor. This was followed in 1954 with the commissioning of the first pilot reprocessing plant at Fontenay-aux-Roses.

An industrial-scale reprocessing plant (Usine de Plutonium 1, UP1) began operating at Marcoule in January 1958 and serviced fuel discharges from three plutonium production reactors on the same site until about 1985, soon after the last of three production reactors was shut-down. The UP1 plant also reprocessed initial fuel discharges with low burn-ups from the three Chinon power reactors in 1965 and 1966.

Reprocessing of power-reactor fuel was switched to a new site—La Hague in Normandy—in the mid-1960s. Here the UP2 plant began operating in 1966, taking fuel from French and Spanish Magnox reactors. Production reached a

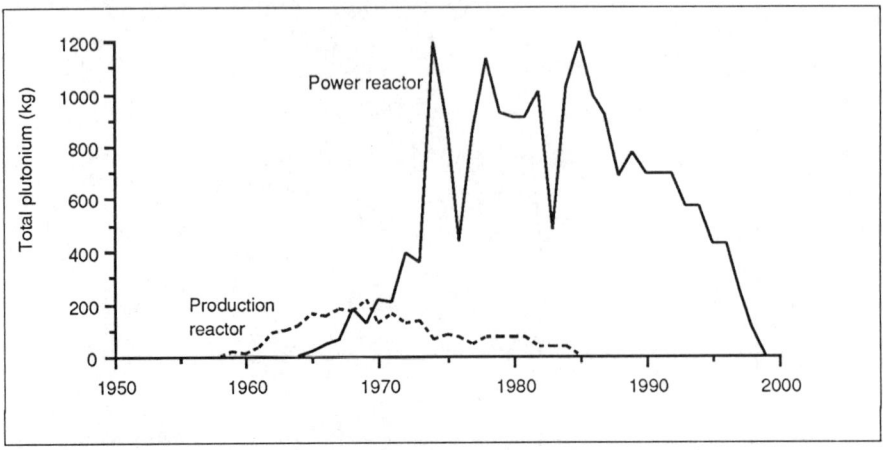

Figure 6.4. Magnox plutonium separation from power-reactor (UP1 and UP2) and plutonium-production reactor fuel in France, 1955–2000

Sources: Syndicat CFDT de l'Énergie Atomique, *Le dossier électronucléaire* (Éditions du Seuil: Paris, 1981), pp. 186–91; Hirsch, H. and Schneider, M., 'Wiederaufarbeitung in Europa: Wackersdorf ist tot es—lebe La Hague?', *Rest-Risiko*, no. 6 (Greenpeace: Hamburg, Apr. 1990).

peak in the mid-1970s before falling and eventually ending in 1987. Meanwhile, reprocessing of Magnox power-reactor fuel restarted at UP1 in the mid-1970s. During the early 1980s all Magnox reprocessing was moved back to Marcoule, leaving facilities at La Hague to concentrate on LWR fuel processing. Co-processing of production reactor and power-reactor fuel ended at UP1 in about 1986. The rate of plutonium separation from magnox fuel at La Hague and Marcoule is shown in figure 6.3.

Magnox fuel reprocessing at UP2 increased gradually over the 10 years from 1965 onwards, reaching a peak of just over 1 tonne of plutonium separated in 1974. Thereafter, with the reintroduction of Marcoule's UP1 for processing civil fuel, UP2 magnox reprocessing fell. By 1987, UP1 was handling all French and Spanish magnox fuel, separating about 700 kg of plutonium annually. French magnox plutonium production was at its peak between 1974 and 1987, when mean annual production was above 900 kg. This is shown in figure 6.4 which contrasts aggregated Magnox power reactor plutonium separation with separation of plutonium from production reactor fuel.

Oxide fuel

Oxide fuel reprocessing began in 1976 at the UP2 plant after a modified head end plant (Haute Activité Oxyde, HAO) was installed. Capacity was gradually

Table 6.5. Existing contracts for reprocessing at La Hague, France, 1991

Figures are in tonnes of heavy metal.

Country	Fuel contracted[a]		
	To end 1989	1990–2000	Post 2000
France	645	7 320	..
Japan	151	2 718	–
Germany	1 725	3 074	2000
Switzerland	132	547	–
Belgium	139	464	..
Netherlands	79	140	..
Sweden	(57)	(727)[b]	..
Total	2 871	14 263	2 000

[a] UP2 contracts are distributed according to the proportion of capacity taken by utilities.

[b] Swedish contracts with Cogema have been exchanged or sold to German utilities (576 t) and Japanese utilities (151 t).

Sources: Cruickshank, A., 'Cogema looks to wider markets', *Nuclear Engineering International*, Sep. 1984, pp. 33–39; MacLachlan, A., 'Swedes transfer reprocessing rights to German utilities', *Nuclear Fuel*, 5 Feb. 1990, pp. 13–14; *Reprocessing News*, no. 15 (United Reprocessors: Hannover, Feb. 1990), p. 3.

raised as magnox fuel processing was phased out at UP2, and reached 400 tonnes per year by the late 1980s. By the end of 1990 almost 3500 tonnes of oxide fuel had been reprocessed at La Hague. Most of this was foreign fuel— less than one-quarter was from French reactors. The contractual situation for reprocessing oxide fuel at La Hague is shown in table 6.5.

During 1990 a new set of facilities (UP3) was brought into operation. It was built to handle fuel from foreign customers under contracts signed after 1978. Once UP3 came into operation, UP2 was to have been dedicated to handling French fuel. In practice, the transition has been less clear-cut, with foreign fuel continuing to be processed through parts of UP2. Capacity at the UP2 plant is planned to be expanded to about 800 tonnes per year, starting in 1994.

UP2 and UP3 are not entirely separate. For instance, solvents from one of the UP3 cells (R2) are sent to UP2 for plutonium and uranium separation.[22] Making too clear a distinction between the two plants may therefore be misleading.

Rates of plutonium separation from oxide fuel at UP2 and UP3 are shown in figure 6.5. In generating these figures, no distinction has been made between BWR and PWR fuel. The figure shows a very gradual buildup of LWR fuel reprocessing at UP2 starting in 1976. Full capacity was reached in 1987 when magnox reprocessing ceased. Today just under 3 tonnes of plutonium are separated annually at UP2. Expansion of the plant will allow a step-jump in production in 1994/95 up to about 6 tonnes a year. Operational experience at

---

[22] Hirsch, H. and Schneider, M., 'Wiederaufarbeitung in Europa: Wackersdorf ist tot es—lebe La Hague?', *Rest-Risiko*, no. 6 (Greenpeace: Hamburg, Apr. 1990), p 10.

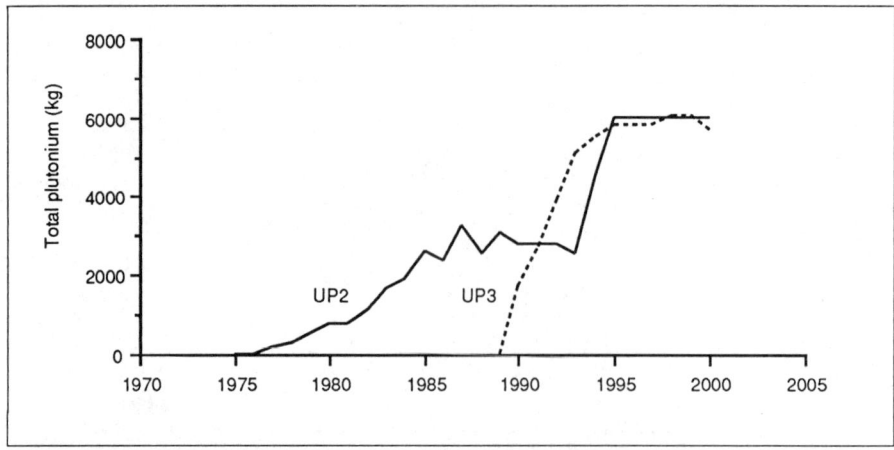

Figure 6.5. Past and projected plutonium separation from LWR fuel at the La Hague UP2 and UP3 reprocessing plants in France, 1975–2002

Sources: Hirsch, H. and Schneider, M., 'Wiederaufarbeitung in Europa: Wackersdorf ist tot es—lebe La Hague?', *Rest-Risiko*, no. 6 (Greenpeace: Hamburg, Apr. 1990); 'Active commissioning of the UP3 plant: the first days . . .', *Reprocessing News*, no. 15 (United Reprocessors Group (URG): Hannover, Feb. 1990); 'UP3 in figures', *Nuclear Engineering International*, June 1992, p. 4.

UP3 was good in the first 24 months of operation, and it is also set to reach full capacity in 1994/95. By then, annual plutonium production at La Hague will be about 12 tonnes.

Fast reactor fuel

Fast reactor fuel reprocessing in France has been carried out in both dedicated plants as well as at UP2. The first dedicated plant was also sited at La Hague, the second and third at Marcoule. These plants have all been on a pilot scale, operated by the CEA.

Laboratory-scale reprocessing of fast reactor fuel began in the Cyrano laboratory at Fontenay-aux-Roses in 1968. Since then about 100 kg of fuel from the Rapsodie and Phénix reactors has been processed there, producing a total of 15–20 kg of plutonium.

The decision to build the Atelier Traitement-1 (AT1) plant at La Hague to reprocess fuel from the Rapsodie fast reactor was made in 1964. AT1 was designed to have the capacity to reprocess one fuel core (134 kg total)[23] per annum and began active operation in 1969 when it processed 220 fuel rods

---

[23] The initial core was 30% enriched with plutonium (about 40 kg). Groupement Centrale Nucléaire Européene à Neutrons Rapides (NERSA), *The Creys-Malville Power Plant*, brochure, EdF, Direction de l'Équipement, Alpe Lyons, 1987, p 8. For a listing of more technical information about French fast reactors, see table 7.1.

weighing 25 kg.[24] Rapsodie was fuelled with both plutonium and HEU. In all, some 910 kg of Rapsodie plutonium fuel was processed at AT1 by the end of 1986.[25] We take this to have been the full amount of MOX discharged from the reactor. Estimating the amount of plutonium contained in this fuel is difficult because initial fuel enrichments are likely to have varied and because of the wide range of burn-ups exhibited in the fuel (up to a maximum of 200 000 MWd/t). However, if we assume that all of the processed fuel initially had a plutonium enrichment of 30 per cent (see table 7.1), and assume a mean burn-up of 20 per cent, then the plutonium inventory in the reprocessed fuel would be about 220 kg Pu_{fiss} (about 290 kg Pu_{tot}).

A small amount of Phénix fuel (about 0.18 tonne) was also reprocessed at AT1. Assuming a mean plutonium enrichment in fresh fuel of 27 per cent (see table 7.1), and a mean burn-up of 20 per cent, this fuel would have contained about 40 kg of Pu_{fiss} (about 50 kg Pu_{tot}). AT1 has now been shut down.

Fast reactor fuel reprocessing at Marcoule began in 1974 when the Service de l'Atelier Pilote (SAP), which had previously treated magnox fuel, was converted to take fuel discharged by Rapsodie. The refurbished plant, Traitement d'Oxydes Pilote (TOP), had a design capacity of 10 kg of fuel per day, although this appears not to have been achieved. Between 1974 and 1976 fuel discharged from Rapsodie (0.05 tonne of MOX), the EL4 heavy-water reactor, and the German KNK1 thermal reactor (an HEU-fuelled reactor, shut down for installation of a fast core in 1974) was reprocessed. In all, 1.65 tonnes of KNK HEU fuel were processed. From 1977 to 1983, when the TOP plant was shut down, only fuel from Phénix was processed there. A total of 9.04 tonnes of Phénix fuel was handled, of which 6.74 tonnes was FBR-MOX fuel. The remainder (2.3 t) was HEU fuel. Assuming 27 per cent enrichment and 20 per cent burn-up, the putonium fuel would have contained about 1450 kg of fissile plutonium (about 1900 kg of plutonium altogether).[26]

In 1979 experiments began with fast reactor fuel reprocessing at UP2. The fuel was declad and dissolved in the HAO facility, and then sent for chemical separation, diluted in magnox fuel solution. A total of 10.07 tonnes of Phénix fuel were treated in this way before magnox reprocessing ended at UP2 in 1987.[27] Assuming a plutonium inventory in spent Phénix fuel of 22 per cent, this would have yielded a total of some 2200 kg of fissile plutonium (about 2900 kg of total plutonium).

The TOP facility (renamed the Traitement d'Oxydes Rapides, TOR) was re-opened in 1988 following refurbishment, with a design throughput of 6 tonnes

---

[24] CEA, 1989 (note 21), p. 164.

[25] Mégy et al. (note 17), p. 285.

[26] This plant is sometimes known by the acronym APM (Atelier Pilote Marcoule). See Davis, M. D., *The Military–Civilian Nuclear Link: a Guide to the French Nuclear Industry* (Westview Press: Boulder, Colo., 1988), pp. 76–77 and p. 104; and Mégy et al. (note 18), p. 285.

[27] CEA, 1982 (note 21), p. 165, states that by 1982, 2.1 t had been processed at UP2. CEA, 1989 (note 21), p. 17, states that of the 20 t of fast reactor fuel reprocessed in France by 1989, 19 t had been discharged by Phénix.

Table 6.6. Cumulative past plutonium separation in France, ends of 1970, 1980 and 1990

Figures are in kilograms of total plutonium.

Reactor type	End of 1970	End of 1980	End of 1990
Magnox	700	7 850	16 550
LWR	..	1 850	25 700
Fast	..	1 190	6 800
Total	**700**	**10 890**	**49 050**

of spent fuel per year. Since then it has processed fuel from the Phénix and KNK II fast reactors at a rate of about 2 tonnes per year. Some LWR MOX fuel has also been processed there.[28] Assuming average plutonium concen-trations in spent Phénix fuel of about 22 per cent, some 1.3 tonnes of fissile plutonium (1.1 t from Phénix fuel and about 90 kg from KNKII fuel) would have been separated at TOR by the end of 1990. In total, therefore, we estimate that about 5.1 tonnes of fissile plutonium (about 6.8 t of total plutonium) have been separated from fast reactor fuel in France. About 90 kg of this material was owned by Germany.

Future production is entirely dependent on the operation of the Phénix and Superphénix reactors (see section IV, chapter 7). At the time of writing both were down following the discovery of a mysterious technical problem involving inert gas bubbles in their sodium coolant and generally critical safety reports by the French nuclear regulator. In July 1992 the French Government refused to issue a new licence for Superphénix, and it seems likely that the reactor will not operate again. If neither reactor operates again but the cores are discharged and reprocessed, this would yield about 6.3 tonnes of plutonium (5.5 tonnes from the Superphénix core and 830 kg from Phénix).

Summary information for plutonium separation in France is presented in table 6.6. In the fast reactor fuel figures we assume that fuel throughput was steady at the four facilities involved (AT1, TOP, UP2 and TOR).

Japan

Magnox fuel

Spent fuel from the single Magnox reactor at Tokai-mura has been sent for reprocessing at Windscale/Sellafield since it started operation. This was under a commercial fuel supply arrangement agreed when the UK sold the reactor, and

---

[28] CEA, *Rapport Annuel 1989* (CEA: Paris, 1989), p. 39. This assumes that about one-third of the Phénix core (1.5 t of fuel) was replaced annually while it was operating and that this was steadily processed. Two cores of MOX fuel weighing 728 kg each, the first enriched to 15% with plutonium and the second to 60%, were loaded at KNK II between 1977 and 1991 when the reactor was closed down. We assume that the first core was reprocessed before 1990 and that the second will be reprocessed some time in the next decade. See *Nucleonics Week*, 5 Sep. 1991, pp. 9–10.

included a provision for separated plutonium to be returned to Japan. The last such shipment took place in 1981, by which time some 800 kg of plutonium had been sent back by BNFL.[29] By the end of 1990 some 1600 kg of plutonium had been separated from Japanese fuel at Sellafield (see table 6.9 below). We can estimate that nearly 1 tonne of Japanese material therefore remains in store in the UK.[30]

Oxide fuel

Reprocessing has long been regarded in Japan as a prerequisite to a strong and independent nuclear power programme. Political and technical obstacles have delayed the establishment of a large capability, but government and utilities are now embarked on an ambitious plan to turn Japan into a major producer and user of plutonium by the beginning of the next century.

Japan's first reprocessing plant for oxide fuel was constructed at Tokai-mura during the 1970s. Although completed in 1974, it did not begin full operation until 1981. The plant's design capacity was set at 210 tonnes per year, but it has operated at well below this level, processing a total of 509.4 tonnes of Japanese PWR, BWR and advanced thermal reactor (ATR) fuel by the end of 1990.[31] By then some 3.6 tonnes of plutonium had been separated (see table 6.7 and figure 6.6). The plant's comparatively poor performance has been put down to technical problems as well as the insistence of utility companies on running separate campaigns for their batches of fuel.

Japanese plutonium production has rather erratically increased to a level of around 600 kg per year. This is comparable, for instance, with French magnox plutonium production. Whereas Germany remained a small producer of plutonium, Japan slowly established itself in the 1980s as having a significant reprocessing capacity, and today has separated more material than any other non-nuclear weapon state (NNWS). It is also the only NNWS party to the NPT with plans to continue separating plutonium.

The Tokai-mura reprocessing plant is unique in that plutonium and uranium are not separated from each other. A plutonium–uranium nitrate solution is used as a direct feed to mixed-oxide fuel fabrication plants on the same site. This was a concession to US concerns in the 1970s that Japan should not acquire a plutonium separation capability.

A large new reprocessing facility is planned to start operation soon after 2000 at Rokkasho-mura in northern Honshu. The plant is being built by the Japan Nuclear Fuel Services Company (JNFS) and is financed by the electric

---

[29] Table 1 in 'Plutonium—do we really need it?', *Nuke-Info Tokyo* (Tokyo) no. 16 (Mar./Apr. 1990), shows that 660 kg of fissile plutonium were returned by the UK to Japan between 1970 and 1981 in a total of 13 shipments, eight of them by air. To derive a total plutonium figure we assume that the material was 83% fissile. The table records just one shipment, of 190 kg Pu_{fiss} from France to Japan, in 1984.

[30] The total amount of foreign plutonium in store at Sellafield has been given as 1 t, although this may include material separated from fuel discharged by the Italian Latina reactor. See British Department of Energy Press Release, 18 Oct. 1990.

[31] 'Tokai marks reprocessing of 500 tonnes of fuel', *PNC Review*, no 17 (spring 1991), p. 6.

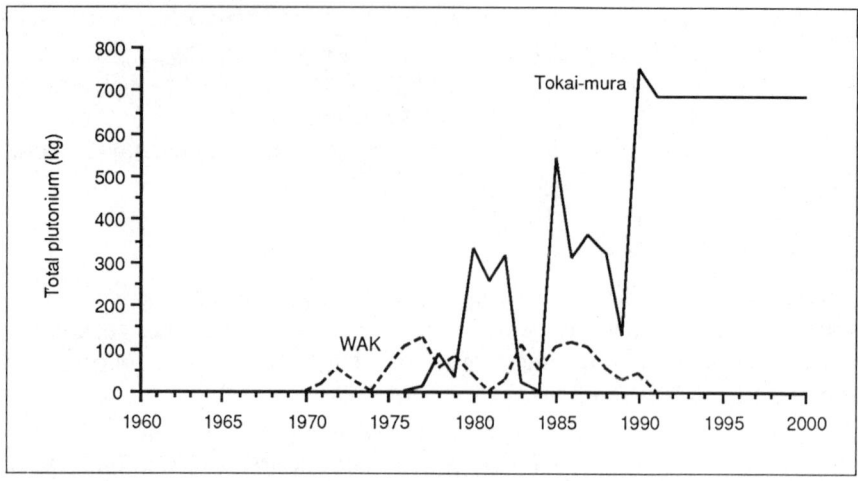

Figure 6.6. Past and projected quantities of plutonium separated at the German WAK and Japanese Tokai-mura reprocessing plants, 1970–2000

Sources: Dr P. Lausch, Deutsche Gesellschaft für Wiederaufarbeitung von Kernbrennstoffe (DWK), Karlsruhe, Private communication with the authors, 20 Jan. 1992; 'Tokai reprocessing plant completes 91-1 campaign', *PNC Review*, no. 19 (autumn 1991), p. 10.

utilities. As with the Tokai plant, it will be based on French technology, and is modelled on the UP3 plant at La Hague. Its design capacity is 800 tonnes of fuel (6–7 tonnes of plutonium) per year.[32] The facility is being built to satisfy only domestic demand. There are as yet no plans for JNFS to become a competitor to BNFL and Cogema in the international fuel services market. Projections of plutonium separation at Rokkasho-mura are given in table 6.11 below.

Germany

Oxide fuel

Reprocessing of oxide fuel began in Germany in 1971. Apart from a number of laboratory rigs, reprocessing was centred at the WAK plant at the Karlsruhe Nuclear Research Centre (KfK). WAK was a pilot facility with a design throughput of 35 tonnes of fuel annually, although this was never achieved because of technical problems. About half of the fuel processed was from German light water reactors, while the rest was discharged by the materials testing heavy water reactor (MZFR) also located at Karlsruhe.[33]

Plans to extend German domestic reprocessing capacity were first unveiled in the early 1970s and went through a number of stages before foundering in 1989 when utility support was withdrawn from a facility being built at

---

[32] Atomic Energy Commission, *Nuclear Fuel Recycling in Japan,* Report by the Advisory Committee on Nuclear Fuel Recycling, Tokyo, 2 Aug. 1991, provisional translation, Aug. 1991.
[33] The MZFR operated between 1966 and 1984.

Wackersdorf in Bavaria. Soon after, a decision was taken jointly by federal and state governments to end funding for reprocessing research, and this led to the closure of WAK on 31 December 1990. A total of 208 tonnes of fuel were reprocessed at WAK, 1164 kg of plutonium being separated.[34]

Figure 6.6 shows the rate of plutonium separation at the WAK and Tokai plants. It shows that German plutonium separation for the 20 years 1971–91 was at a relatively low level, never going above 13 kg per year.

The United States

Besides the military plutonium separation facilities at Hanford in Washington State and the Savannah River site in South Carolina (see chapter 3), three commercial reprocessing plants were built in the USA: West Valley, Morris and Barnwell.

Only one, the Nuclear Fuel Services plant at West Valley, New York, ever operated. This was a PUREX process plant with a design capacity of 300 tonnes of oxide fuel per year. West Valley operated between 1966 and 1972. During this time, 2058 kg of plutonium were separated from the 676 tonnes of fuel reprocessed. Over half of the fuel reprocessed (375 tonnes) came from the N-reactor at Hanford, although this had a low burn-up, yielding a total of only 553 kg of plutonium. A total of 1505 kg was separated from 301 tonnes of commercial power-reactor fuel.

The Midwest reprocessing plant at Morris, Illinois, built by General Electric with an annual capacity of 300 tonnes of fuel, ran into serious technical difficulties during cold testing and was declared inoperable in 1974. GE subsequently abandoned the project.[35] Beginning in the late 1960s, Allied-General Nuclear Services (AGNS) built a reprocessing facility at Barnwell, adjacent to the Department of Energy Savannah River site in South Carolina. It was due to begin operation in 1974 with a nominal capacity of 1500 tonnes per year. Following delays in construction and licensing, it had still not been finished in 1977 when President Carter decided to defer indefinitely all commercial reprocessing. Federal funding was cut, and licensing and construction suspended. The Barnwell facility was also eventually abandoned and the plant decommissioned.

No power-reactor fuel has therefore been reprocessed in the USA since the West Valley plant closed in 1972.

The former Soviet Union

The decision to reprocess spent fuel from power reactors was not taken in the Soviet Union until 1978, 14 years after the first power reactor (Kursk 1) began

---

[34] Personal communications with Dr P. Lausch, Deutsche Gesellschaft für Wiederaufbeitung von Kernbrennstoffe (DWK), Karlsruhe, Jan. 1992.

[35] Rochlin, G. I. (note 11), p. 72.

operation. Even then, only fuel from the VVER-400 series reactor was handled in this way.[36] RBMK fuel is not reprocessed. Before 1978, fuel was stored predominantly at reactor sites, although some VVER-100 fuel was also sent to a centralized store at Chelyabinsk. The commitment to reprocessing was underlined in the 'Long-Term Energy Programme' agreed by the Soviet Council of Ministers in 1984.

All these activities are concentrated at the Radiochemical Combine's Mayak plant in Chelyabinsk-40. The reprocessing plant (RT-1) began operation in 1956 and operated for 20 years reprocessing production reactor fuel. It was modified in 1976 to take higher burn-up fuel from power reactors and naval propulsion reactors. Since then it has reprocessed fuel from 23 reactors, eight in the Soviet Union, the rest in Eastern Europe and Finland.[37] The plant's capacity is put at 600 tonnes or 300–900 fuel assemblies per year, although its current capacity is between 200 and 300 tonnes per year.

It has been variously estimated that 20 or 25 tonnes of plutonium had accumulated at Chelyabinsk-40 by the end of 1990.[38] However, it is not clear whether all of this material was separated from power-reactor fuel, or whether it includes material from other sources such as submarine reactor fuel. Assuming that 80 per cent of the plutonium is separated from power-reactor fuel, and further assuming a mean plutonium inventory of about 8 kg/tonne of fuel, then it can be estimated that achieved throughputs at Mayak have averaged between 170 and 210 tonnes of fuel per year.

India

India is the only country committed to reprocessing CANDU power-reactor fuel. In late 1982, the Power Reactor Fuel Reprocessing (PREFRE) plant, located at the Bhabha Atomic Research Centre (BARC) at Tarapur near Bombay, started processing safeguarded natural uranium spent fuel from the Rajasthan Atomic Power Station 1 (RAPS-1, CANDU).[39] By late 1983, it had reprocessed about 20 tonnes of RAPS spent fuel, much of which appeared to have had a low burn-up since it was reported to have contained only 25 kg of plutonium.[40] According to the IAEA, PREFRE has not processed any additional RAPS spent fuel since then. In late 1985 or early 1986, the PREFRE facility

[36] Although a facility dedicated for VVER-1000 fuel at Krasnoyarsk is 30% complete, the project was abandoned in the late 1980s because of local opposition. See 'A visit to Shary Shagan and Kyshtym', *Science and Global Security,* Occasional Report, vol. 1, nos 1–2 (1989), p. 173.

[37] Bukharin, O., 'Soviet reprocessing and waste management strategies', paper presented to the International Workshop on Nuclear Weapons Disposal, Berlin, 28–30 Nov. 1991.

[38] Bukharin (note 37); *Science and Global Security* (note 36). By mid-1992 the total separated was about 30 t (personal communication, Evgeny Mikerin, 18 May 1992).

[39] 'India reprocesses RAPS fuel under IAEA eyes', *Nuclear Fuel,* 14 Feb. 1983, p. 12.

[40] Abraham, A., 'Plutonium missing in Tarapur plant', *Sunday Observer* (Bombay, India), 16–22 Oct. 1983.

began reprocessing spent fuel from the unsafeguarded Madras Atomic Power Station (MAPS).[41]

The nominal daily reprocessing capacity of the PREFRE facility is 0.5 tonne of uranium.[42] Assuming the plant operates about 50 per cent of the time, its nominal annual capacity is 100 tonnes of CANDU spent fuel a year. The Chairman of the Indian Atomic Energy Commission said in late 1990 that the capacity of the facility had been increased to a maximum of 150 tonnes per year.[43] If the CANDU spent fuel has nominal burn-ups of 6700 MWd/t, it would contain about 3.5 kg of plutonium per tonne of fuel. Therefore, at these capacities, PREFRE could theoretically separate 350–525 kg of reactor-grade plutonium a year.

India is building a second CANDU reprocessing plant at the Indira Gandhi Centre for Atomic Research in Kalpakkam near Madras. This facility is also scheduled to process spent fast reactor fuel from the 40-MWth test fast reactor at Kalpakkam. Originally intended for completion in 1990, the Kalpakkam reprocessing plant will not now be ready for hot operation until 1993 or 1994. India originally stated that this facility would have a capacity of 100 tonnes of spent CANDU fuel a year, but it now claims that throughput could reach over 200 tonnes a year, for an output of about 700 kg of plutonium per year.[44]

Since PREFRE has processed only unsafeguarded fuel since 1985, its throughput up to 1990 was limited by the amount of plutonium produced in the two MAPS reactors. These reactors have not operated as well as expected, showing a mean cumulative load factor of only 47 per cent at the end of 1991.[45] Assuming an optimal fuel-management regime, we estimate that these two reactors together discharged annually about 40 tonnes of fuel, or a total of 200 tonnes to the end of 1990. Allowing two years to cool the fuel sufficiently for handling and transportation, PREFRE could have processed about 120 tonnes of fuel by then. Most of this fuel (112 tonnes) would have been the first cores. In the first few years of operation a reactor discharges relatively low burn-up fuel. The first cores can be shown to have a mean burn-up about one-half of that achieved when fuel is discharged at equilibrium (see appendix B). Therefore the mean burn-up in the MAPS first cores would have been about 3800 MWd/t, yielding some 2.3 kg total plutonium per tonne. The rest of the fuel (8 tonnes) we estimate to have been discharged at a nominal burn-up of 6700 MWd/t with a plutonium inventory of 3.5 kg per tonne. Overall, therefore, we estimate that, by the end of 1990, some 300 kg of plutonium could have been separated from MAPS fuel at PREFRE.

[41] 'India's supply of unsafeguarded Pu grows as reprocessing of MAPS fuel begins', *Nuclear Fuel*, 11 Aug. 1986.

[42] 'India's significant efforts on reprocessing and vitrification', *Nuclear Europe*, Jan. 1983, pp. 45–46.

[43] Hibbs, M., 'Indian reprocessing program grows, increasing stock of unsafeguarded Pu', *Nuclear Fuel*, 15 Oct. 1990.

[44] Interview with Dr P. K. Iyengar, Bombay, 5 Sep. 1990.

[45] *Nuclear Engineering International* (note 19), p. 21; interview with Dr A. N. Prasad, Bombay, Mar. 1992.

Not all of this MAPS plutonium, however, is likely to have been separated. According to Dr A. N. Prasad, Director of Fuel Reprocessing and Waste Management Group, Department of Atomic Energy, PREFRE has only separated plutonium for research purposes, which include a reload of the test breeder reactor whose core contains about 80 kg of plutonium. We therefore estimate that PREFRE has separated in total no more than about 100 to 200 kg of plutonium from MAPS and RAPS fuel.

Estimating the future growth in the civilian plutonium inventory is difficult, since so many uncertainties surround the question of when new reactors will start and when the Kalpakkam reprocessing plant will operate. At minimum, PREFRE could be expected to separate about 50 to 500 kg of plutonium a year, giving a total of 500 to 1500 kg separated during the 1990s.

India wants to accumulate a stockpile of at least 2000 kg of plutonium for the initial core-load of a first 500-MWe fast reactor which it plans to start by 2005.[46] In addition to the first core load of plutonium, Indian officials estimate they will require another 3000 kg to refuel the reactor during the first five years of operation. After that, they plan to recover plutonium from the breeder spent fuel.

If India is to achieve a goal of roughly 5000 kg of plutonium by the year 2005, it will have to depend almost totally on the Kalpakkam Plant. It could do this by processing an average of about 150 tonnes a year at Kalpakkam from 1995 to the end of 2005.

Argentina

In 1978, Argentina announced its decision to build a small reprocessing facility at the Ezeiza Research Complex near Buenos Aires. When completed, it is expected to separate about 15 kg of plutonium a year from about 5 tonnes of heavy water spent fuel discharged from its Atucha 1 heavy water reactor.[47] The facility has suffered many delays, and is still not fully operational. Argentina has stated plans to fabricate the plutonium into fuel for the Atucha 1 reactor and is building a small plutonium fuel fabrication plant next to the reprocessing facility.

The National Atomic Energy Commission (CNEA) wants to build a larger reprocessing plant at the end of this century. The CNEA has not issued any concrete plans for this facility, but a CNEA official in charge of the plutonium fuel programme said in August 1988 that the facility would be able to reprocess about 80–90 tonnes of spent fuel a year, recovering about 240–70 kg of plutonium a year.[48] He said that this projected capacity corresponds to the amount of plutonium that can be used in the two Atucha reactors.

---

[46] Interview with Dr P. K. Iyengar, Bombay, 18 Mar. 1992.

[47] 'Argentina's CNEA decides to reduce scope of near-term fuel reprocessing program', *Nuclear Fuel,* 12 Aug. 1985.

[48] Interview in Buenos Aires, Aug. 1988.

Table 6.7. Cumulative separation of plutonium from power and research reactor fuel, ends of 1980, 1990 and 2000

Figures are in kilograms of total plutonium.

Country	Plant/ fuel	End of 1980	End of 1990	End of 2000 (current plans[a])
Belgium	Eurochemic (LWR)	680	680	680
France	UP1 (civil)	1 500	7 800	11 500
	UP2 (Magnox)	6 400	8 700	8 800
	UP2 (LWR)	1 900	24 000	72 800
	UP2[b] (FBR)	(–)	(2 900)	(2 900)
	UP3 (LWR)	–	1 700	54 200
	Marcoule[b] (FBR)	(1 000)	(3 600)	(300)
	AT1[b] (FBR)	(200)	(300)	
FRG	WAK (LWR)	530	1 165	1 165
Japan	Tokai (LWR)	440	3 600	10 300
UK	B205 (Magnox)	23 200	46 650	66 700
	B204/205 (LWR)	360	360	360
	THORP (LWR/AGR)	–	–	36 300[c]
	DNPDE[b] (FBR/Magnox)		(3 000)	(5 530)
USA	West Valley (LWR)	1 505	1 505	1 505
Subtotal		**36 515**	**96 160**	**264 310**
Russia	Chelyabinsk (VVER)	3 000	25 000	25 000–40 000
India	Tarapur (CANDU)	0	100–200	500–1 500
	Kalpakkam (CANDU)	–	–	1 500–3 000
Total		**39 515**	**121 260**	**290–310 000**

[a] For a discussion of future reprocessing scenarios, see section VII in this chapter.

[b] Materials from fast-reactor fuel separated at UP2, Marcoule, AT1 and DNPDE fast-reactor fuel facilities are not included in the totals because they are *recycled* plutonium and therefore do not add to total stocks.

[c] A total of 47 200 kg is due to be separated at THORP by 2002 when the first tranche of contracts is completed.

VI. Summary of past power- and fast-reactor fuel reprocessing

By the end of 1990 some 120 tonnes of plutonium had been separated from power-reactor fuel at plants around the world, a rise from 40 tonnes in 1980. Separated plutonium represents less than one-fifth of the plutonium discharged from power reactors to date (the remainder being held in stored spent fuels). Summary information on cumulative plutonium separation at individual facilities is given in table 6.7.

This table contains data of variable quality, the best being from facilities which have ceased operating, such as Eurochemic, WAK and West Valley, the

Table 6.8. Cumulative distribution of plutonium separation from power-reactor fuel, ends of 1980, 1990 and 2000

Figures are percentages.

Country	End of 1980	End of 1990	End of 2000
France	23	35	47
UK	55	39	33
USSR/Russia	16	20	14
Others	6	6	6
Total	**100**	**100**	**100**

worst being from the Tarapur and Chelyabinsk facilities for which no published operating information is currently available. Information for the UK, France, Japan and Germany is from a variety of sources and includes some modelling assumptions. These are spelt out in appendix C. The totals should therefore be regarded as central estimates, with the historical sub-totals given for Europe, the USA and Japan (36.5 tonnes by 1980 and 96 tonnes by 1990) being more reliable than the final, world total.

Figures for the year 2000 are also given in table 6.7 and these are based on the information given about reprocessing contracts in the previous section. These represent an indicative *maximum* estimate of plutonium separation. A more detailed discussion of possible future scenarios is given in section VII.

Plutonium separation has been a highly concentrated industrial activity. The great majority of commercial reprocessing has occurred in NWS and at only a few sites. Three sites—Windscale/Sellafield (UK), La Hague (France), and Chelyabinsk (Russia)—account for over 87 per cent of the plutonium separated from power-reactor fuel to date. This concentration will increase over the coming decade. Table 6.7 suggests that over 85 per cent of cumulative plutonium separation will have taken place at these sites by 2000. Only about 4.5 per cent of plutonium separation has so far taken place in NNWS (in Japan, Germany and Belgium). To a great extent, nuclear weapon states have maintained their domination over the production of plutonium. This has been due as much to the lack of a commercial value for plutonium as to active policies of dissuasion by NWS, and difficulties in licensing reprocessing plants in NNWS.

Within the NWS there have been shifts in the relative proportions of plutonium separation which each has undertaken (see table 6.8). Up to the early 1980s over half of world cumulative separation from power-reactor fuel had taken place at Windscale/Sellafield. With the start of large-scale oxide fuel reprocessing at La Hague, the relative amount of plutonium separated in France has risen, mainly at the expense of the UK. This trend is set to continue. If current plans are implemented, some 47 per cent of cumulative separation will have taken place in France by 2000.

Table 6.9. Cumulative plutonium separated in the UK and France from fuel from non-nuclear weapon states, ends of 1980, 1990 and 2000

Figures are in kilograms of total plutonium.

	Past separation		Projected separation	
	End of 1980	End of 1990	End of 2000[a]	By 2002
Plutonium separated from NNWS fuel in the UK [b]				
FRG	..	..	7 000	(8 900)
Italy	1 400	2 600	4 500	(4 700)
Japan	800	1 600	21 000	(26 500)
Netherlands	..	..	500	(500)
Spain	..	..	1 000	(1 600)
Switzerland	..	..	3 000	(4 200)
UK total	**2 200**	**4 200**	**37 000**[c]	**(46 400)**[c]
Plutonium separated from NNWS fuel in France				
Belgium	..	1 170	5 000	
FRG	1 100[d]	14 580	40 000	
Japan	..	1 170	23 500	
Netherlands	..	670	1 500	
Switzerland	..	1 110	5 500	
France total	**1 100**	**18 700**	**75 500**	
Total	**3 300**	**22 900**	**112 500**	

[a] Assumes steady reprocessing of all foreign fuel and that the mean fuel burn-ups are equal.
[b] Not including B204/205 separated material, amounting to around 300 kg.
[c] Of this, 5300 kg will have been separated from magnox fuel.
[d] Assumes that 60% of LWR fuel reprocessed at La Hague by the end of 1980 was German (see appendix C).

Sources: Berkhout, F. and Walker, W., 'Spent fuel and plutonium policies in Western Europe: the non-nuclear weapon states', *Energy Policy*, July/Aug. 1991, pp. 553–68; Albright, D., *World Inventories of Plutonium*, PU/CEES Report no. 195 (Center for Energy and Environmental Studies, Princeton University: Princeton, N. J., June 1987), table 8, p. 41; authors' estimates. (For basic data, see tables 6.4, 6.5 and 6.7.)

Ownership of plutonium

While plutonium separation has been carried out mainly in three NWS, the ownership of this material is more dispersed. Plutonium separated from foreign fuel in the USSR was not returned under arrangements which held until 1990. Since then the reprocessing industry in Russia has been placed on a commercial footing, with the option of plutonium return. However, it is unclear which former Soviet dependencies will continue to send fuel to Russia, and whether any of them will wish to take possession of separated plutonium.

Table 6.10. Cumulative discharged power-reactor plutonium which has been separated, ends of 1980, 1990 and 2000

	End of 1980	End of 1990	End of 2000
Plutonium discharged (tonnes)	144	654	1 390
Plutonium separated (tonnes)	40	121	310
Percentage separated	28	19	22
Percentage separated from NNWS fuel	10	25	42

In the West, plutonium generally remains the property of the utility in whose reactor it was produced.[49] Both the UK and France have already reprocessed substantial amounts of fuel from other countries, and they will expand this activity over the coming decade. Fuel from Magnox reactors in Japan and Italy has been routinely reprocessed at Sellafield and Marcoule since the 1960s. Details on this activity are sparse, but it is assumed here that all the fuel from the Tokai and Latina reactors is reprocessed two years after discharge. We therefore estimate that by the end of 1990 some 2400 tonnes of foreign magnox spent fuel had been reprocessed at Sellafield, and that this will rise to almost 3200 tones by 2000, when it is expected to end following the closure of the Tokai reactor.

LWR fuel has also been reprocessed from foreign reactors at both Sellafield and La Hague. The British operation lasted just three years and produced about one-third of a tonne of plutonium (see section V above), but oxide reprocessing will expand when the THORP plant begins operation in late 1992. Oxide fuel reprocessing at La Hague has been on a much larger scale, some 60 per cent of which has been German fuel (see table 6.5).

In table 6.9, estimates are set out of the amount of plutonium separated from fuel sent by NNWS customers of French and British reprocessors. These estimates include published data, information received in correspondence, as well as spent fuel discharge figures derived for Japanese and Italian Magnox reactors. The projected figures are based on a pro rata allocation of plutonium separated at facilities according to the size of contracts held by utilities in each country. These are given in tables 6.4 and 6.5.

Table 6.9 shows that the main NNWS customers for reprocessing services in France and the UK are Germany and Japan. By early next century, utility companies in these countries are expected to own 47 and 44 tonnes of separated plutonium respectively, a proportion of which will have been recycled in plutonium fuel. Not all of this material will have been returned by then, but the rights to decide whether to store or to use the material will rest with the owner utilities.

Tables 6.7 and 6.9 also shows that an increasing proportion of the plutonium separated in the UK and France has belonged or will belong to utilities in the

---

[49] The exception is Spanish magnox plutonium separated at Marcoule (see the discussion on France above).

NNWS: 8 per cent in 1980; 19 per cent in 1990; and 35 per cent by 2000. If the material separated in the NNWS is added, then by 2000 about two-fifths of the world's plutonium separated from commercial fuels will belong to the NNWS.

The global significance of reprocessing as a plutonium management strategy

As a proportion of the total plutonium discharged from power reactors, plutonium separation has fallen over the past decade, although it will probably climb again during the 1990s. This is mainly the result of the low level of oxide fuel reprocessing in the past and the expansion of oxide reprocessing over the coming decade. Gross production and separation figures are presented in table 6.10. This table also shows the growing proportion of separated plutonium that belongs to non-nuclear weapon states.

Despite the great increase in reprocessing capacity, most plutonium (75–80 per cent) is set to remain in spent fuel. Without significant changes in the price of uranium, enrichment and reprocessing, even this relatively small amount of material will largely be thought of as surplus to requirements. In terms of quantities, plutonium held in spent fuel is likely to remain the single largest inventory, rising from about 530 tonnes (653 less 120 tonnes) in 1992 to over 1000 tonnes by the year 2000.

VII. Projections of plutonium separation to 2010

In this section three scenarios (maximum, stretch-out and phase-out) of plutonium separation from civil fuels are presented. Describing a maximum forecast of plutonium separation for the next decade (that is, 1992–2001) is relatively straightforward. Assuming the new British and French reprocessing capacity comes on-line, the stock of reprocessing plant will be fixed until the next century. Magnox lines can be expected to continue operating as they have done for the past 25 to 30 years, while oxide fuel reprocessing will take place at Tokai-mura, Sellafield and La Hague on the basis of well-established commercial arrangements. Even if there is some doubt about whether the British and French plants will operate precisely to plan, the perception is that fuel reprocessing is now a mature technology in which the risks of major technical problems have been curtailed. Oxide reprocessing may also continue at Chelyabinsk.

For the post-2000 period, it is less easy to decide on a maximum scenario. Some contracts have been signed for work at THORP and La Hague, but it is uncertain how much of the capacity at these plants will be contracted for. For instance, it is unlikely that Japanese utilities, such important customers during the 1990s, will sign new contracts in France and the UK if they persevere with their plans to develop Japan's own reprocessing capacity. On the other hand,

Table 6.11. Projected mean annual fuel throughputs at commercial reprocessing plants, 1991–2000 and 2001–10: three scenarios

Figures are tonnes of fuel per year.[a]

Facility	1991–2000			2001–10		
	Phase-out	Stretch-out	Maximum	Phase-out	Stretch-out	Maximum
B205	850	850	850	–	170	340
THORP	–	300	650	–	300	700
UP1	250	250	250	–	–	–
UP2-800	–	400	800	–	400	800
UP3	200	500	800	–	310	800
Tokai	50	70	90	–	–	–
Rokkasho	–	–	–	–	400	800
Mayak	–	100	250	–	–	–
Total (magnox)	1 100	1 100	1 100	–	170	340
Total (oxide)	250	1 370	2 590	–	1 410	3 100

[a] Dates of first operation of reprocessing plants are given in table 6.2. It is assumed that all plants take three years to reach full capacity after start-up.

new customers such as South Korea could enter the frame. Nevertheless, by assuming that all the capacity will be filled, and that the Rokkasho-mura plant will be constructed according to plan, it is possible to put an upper bound on plutonium separation up to 2010. There appears to be little likelihood that any facilities not already planned will be built.

Maximum scenarios have a habit of not being realized. We have also made estimates for two alternative scenarios: phase-out and stretch-out. The phase-out scenario envisages a concerted international effort to stem plutonium production, together with a collapse of confidence in the economic and political viability of reprocessing and plutonium recycle. Under the phase-out scenario we assume that THORP and UP2-800 do not operate, and that the Mayak plant at Chelyabinsk is shut down. We further assume that UP3 operates at one-quarter capacity (200 tonnes/year), and that Tokai-mura operates at half its current capacity (50 tonnes/year). Only B205 and UP1 continue operating normally, although there is a subsidiary assumption that all the British magnox reactors are decommissioned by 2000.

In the stretch-out scenario, political and economic pressures would also bring a reduction in the rate of reprocessing, but not to the same drastic extent. As shown in chapter 7, this scenario might be consistent with a policy of minimizing plutonium surpluses while maintaining active MOX programmes. Under stretch-out, oxide reprocessing contracts due to be completed in the period 1990–2002 would be extended over 20 years to 2010. This would approximately halve the throughput at the four large reprocessing plants in the UK, France and Japan. Again, magnox reprocessing would continue as before, with decommissioning of the last British Magnox plant now delayed until 2005; and the Mayak plant would not operate after 2000.

Table 6.12. Spent fuel arisings and plutonium separation, 1991–2000 and 2001–10: three scenarios

Figures are in tonnes.[a]

	1991–2000	2001–10
Spent fuel (oxide)		
Phase-out	2 740	–
Stretch-out	12 400	14 600
Maximum	21 830	27 300
Plutonium (oxide)		
Phase-out	22	–
Stretch-out	99	124
Maximum	164	231
Plutonium (magnox)		
Phase-out	22	–
Stretch-out	23	2
Maximum	24	5
Total plutonium		
Phase-out	44	–
Stretch-out	122	126
Maximum	188	236

[a] Pu separation totals are derived from fuel throughput assumptions set out in table 6.11. The Pu-to-fuel conversion factors for the 'phase-out' and 'stretch-out' scenarios in 1991–2000 are based on a simple assumption of 8 kg Pu/t of fuel. Figures for the 'maximum' scenario are based on the Pu separation figures given in table 6.7. Mean Pu concentration in this fuel is assumed to be lower, at about 7.5 kg/t of fuel. For higher burn-up oxide fuel reprocessed after 2001, the Pu concentration is assumed to be 8.5 kg/t.

The mean plant throughputs under these three scenarios are set out in table 6.11. Estimates of the amounts of plutonium separated are presented in table 6.12. As one would expect, the range of possible outcomes is very large. If the nuclear industry's present plans are carried through to the fullest extent (the maximum scenario), around 420 tonnes of plutonium would be separated from power-reactor fuel between 1992 and 2010. If reprocessing schedules were stretched out, the amount might be closer to 250 tonnes. Were there to be a decisive move away from reprocessing, the quantities of plutonium produced over this time-scale would obviously be very much smaller, only 44 tonnes.

VIII. Conclusion

We estimate that over the next decade between 170 and 190 tonnes of plutonium will be separated from power-reactor fuel. This compares with the 120 tonnes which had been separated by the end of 1990. Reprocessing will be carried out at 11 plants in five countries, four of which are parties to the NPT. The reprocessing of power-reactor fuel is a large-scale and complex industrial undertaking which today is geographically highly concentrated. To date, nearly nine-tenths of all civil plutonium separation has taken place in nuclear weapon states. Until Japan begins operating its planned reprocessing facility at Rokkasho-mura, this geographical concentration is likely to become even more marked. However, the ownership of the plutonium separated by reprocessing is much more diffuse. As a result, substantial international flows of separated plutonium and of plutonium fuel will develop during the 1990s.

Having estimated how much plutonium has been produced and separated, in the next chapter the authors consider how plutonium has been used, where stocks are to be found, and how widely plutonium fuels are likely to be used in the future. This allows some broad conclusions to be drawn about the future balance of supply and demand.

7. Commercial and research and development uses of plutonium

I. Introduction

By the end of 1991 some 130 tonnes of plutonium had been separated from power reactor fuel around the world. This was approximately half the amount separated for weapon programmes. It is shown in chapter 6 that the great majority of civil reprocessing has been carried out in nuclear weapon states (NWS), and specifically in the UK, France and the former Soviet Union, although in the future a growing proportion of the plutonium separated there will belong to non-nuclear weapon states (NNWS). The final part of this assessment of civil plutonium considers how much separated material has been used and where, and goes on to estimate the future balance between supply and demand.

Plutonium was seen, from the beginning of the nuclear enterprise, as a valuable fuel. It was also recognized that the most effective way of using the material would be in fast reactors, rather than in thermal reactors. In some of the more ambitious plans laid for nuclear power in the 1970s, large programmes of fast reactors would lead to the gradual replacement of thermal reactors, the latter providing the initial plutonium to fuel the fast reactors.[1] With this a more or less self-sustaining 'plutonium economy' could be established, plutonium being both created and burned in fast reactors. Continuous reprocessing of nuclear fuels would be an integral part of the fuel cycle.

Such plans were never realized, and the prospects for commercial fast-reactor programmes are, for the present, dim. By the early 1980s it was clear that if power-reactor fuel reprocessing was to make any sense, a new use would have to be found for plutonium. This led several countries committed to reprocessing to develop programmes for large-scale plutonium recycling in conventional thermal reactors. Although a less effective way of using plutonium (see chapter 2, section III), this policy did at least create a demand for a material which the NNWS, in particular, were loathe to stockpile for political reasons. In coming years, thermal recycle will be the main disposal route for separated plutonium.

This chapter is primarily concerned with the use of plutonium separated from civil fuels. Sections II–IV deal with fast-reactor plutonium fuel. Sections V and

---

[1] An influential report came from the World Energy Conference, *Nuclear Resources: The Full Reports to the Conservation Commission of the World Energy Conference* (IPC Science and Technology Press: Guildford/New York, 1978).

VI are concerned with the recycling of plutonium in thermal reactors. There has been considerable speculation recently that plutonium extracted from dismantled nuclear weapons might be used to fuel power reactors. This issue is considered in chapters 2, 12 and 13.

II. Fast-reactor fuel cycles

Fast reactors may be fuelled with either HEU or plutonium. Prototypes have often been fuelled with HEU, but the expectation has always been that plutonium would become the principal fuel for commercial fast reactors. Because fission cross-sections are much lower in fast than in thermal reactors (see chapter 2), fuel must contain higher concentrations of fissile material. The proportions of fissile uranium or plutonium contained in fast-reactor fuel therefore range from 15 to 30 per cent rather than 3 to 5 per cent in thermal reactors. Fast reactor cores are also less uniform than thermal reactor cores, often being surrounded by arrays of depleted or fresh uranium fuel elements (blankets) in which plutonium is produced by neutron capture. This may be recovered for recycling.

The reprocessing of fast-reactor fuels involves the same technology as thermal reactor fuel reprocessing, with some modifications to take account of the greater plutonium inventories and the more intense radioactivity produced at higher fuel burn-ups. Fast-reactor fuel reprocessing capacities are reviewed in chapter 6.

When used as a fuel, plutonium is blended with natural or depleted uranium to form a mixed oxide (MOX). The majority of the fissile worth of the fuel is then provided by the plutonium. This mixture can be sintered into fuel pellets and inserted into a steel or zircalloy cladding. Handling plutonium is more hazardous than handling uranium because of its radioactivity and toxicity. Inhalation must be avoided, and extensive radiation shielding, which is not necessary with uranium, is also usually required. These precautions lead to a more capital-intensive and costly fabrication process for MOX fuel.

The rate at which separated plutonium oxide is made into fuel is to a large extent determined by how much plutonium-241 it contains. After the last purification step at the reprocessing plant, emissions of gamma radiation build up as a result of the gradual accumulation of americium-241 produced by the decay of ^{241}Pu. This leads to progressively more severe radiological constraints on plutonium handling, particularly with plutonium derived from high burn-up LWR fuels which contain relatively high concentrations of ^{241}Pu (see table 2.1). At currently-operating MOX fabrication plants, LWR plutonium cannot be stored for longer than three years before it becomes too hazardous to handle. Once the plutonium has been manufactured into fuel rods it can be stored safely for much longer, usually over a decade.

This section is primarily concerned with fuel in the core of fast reactors (usually divided into an outer and inner core with different enrichments). Blanket fuel is not considered here.

In a typical fast reactor operating to design capacities, between 30 and 40 per cent of the fuel core is replaced each year. Once discharged, the fuel is stored and allowed to cool, either under water or (as at Superphénix) in sodium. The fuel may then be reprocessed. Fast-reactor fuel cycles were originally planned to be independent and self-sustaining, but this has not been achieved anywhere yet. In most cases fast reactors have been fuelled with plutonium separated from thermal fuels. Only in the UK and France has fast-reactor fuel been consistently reprocessed and its plutonium recycled.

III. Plutonium use in fast reactors

To date, 17 fast reactors have been built in eight countries (counting the new CIS states separately). They are all either prototype or demonstration units, with only a single commercial-scale reactor—Superphénix at Creys-Malville in France. The full list of fast reactors, together with basic fuelling information, is given in table 7.1. Operating and planned plutonium fuel fabrication plant capacities are shown in table 7.2.

The amounts given in table 7.1 refer to fissile rather than total plutonium. This is because data on reactor cores are usually published in this form, without details being given about the origin of the material. In general, fast reactors in Europe and the USA had first cores containing reactor-grade plutonium when they were not fuelled with HEU. Some plutonium was low burn-up material discharged from Magnox reactors (about 80 per cent fissile), some was separated from LWR fuel (about 70 per cent fissile), while in France and the UK some fast-reactor core plutonium has also been recycled. To avoid the complications introduced by estimating the fissile content, fissile amounts (Pu_{fiss}) are used in calculating plutonium in initial cores, together with estimated total plutonium figures (Pu_{tot}) where necessary. The general assumption is that plutonium used to fuel fast reactors is 75 per cent fissile.

According to the data in table 7.1, about 10 tonnes of fissile plutonium were needed to fabricate the first cores for the eight reactors fuelled initially with plutonium. In theory refuelling with plutonium could create a demand of about 3.5 tonnes per year. This is currently a considerable over-estimate. Fast reactors have, on the whole, operated erratically. Most have been experimental facilities not intended to operate as profitable electricity producers, and several have encountered technical problems which have led to their shutdown.

IV. Past and projected plutonium consumption in fast reactors

Published information on fast-reactor fuel cycles is rather scarce. However, there are relatively few operating reactors and their basic characteristics are published in open literature. It is therefore possible to go some way towards reconstructing fuelling histories by taking account of their operating perfor-

Table 7.1. The world's fast reactors

Country	Reactor	Power (MWth)	Operated	Fuel	Initial core (THM)	Initial core (kg Pu$_{fiss}$)	Maximum reload (kg Pu$_{fiss}$)
Shut down or abandoned							
USA	EBR1	1–2	1951–63	HEU	?	?	?
Russia	Obinsk BR5	5–10	1954–59	Pu + U	?	?	?
UK	Dourreay DFR	60	1962–72	HEU	0.34	0	0
USA	Fermi 1	300	1966–72	HEU	2.6	0	0
France	Rapsodie	20	1967–83	HEU/Pu	0.040	0	0[a]
Germany	KNK-2	~100	1977–91	HEU/Pu	0.73	430	0[a]
Germany	Kalkar SNR300	–		Pu + U	5.1	1 400	350–400
Operational							
USA	EBR2/1FR	62.5	1963–	HEU	0.46	0	0
Russia	BOR60	60	1969–	HEU	0.176	0	Experimented with Pu
Kazakhstan	BN350	1 000	1972–	HEU/Pu	1.17	0	Experimented with Pu
France	Phénix	560	1973–	HEU/Pu	4.3	830	400–500
UK	Dourreay PFR	600	1974–	Pu + U	4.1	1 100	400–500
Japan	Joyo	100	1977–	Pu + U	0.76	0[b]	60–120
Russia	BN600	1 470	1980–	HEU/Pu	1.26	0	Experimented with Pu
USA	FFTF	400	1980–	Pu + U	1.87	640	200
USA	ZPPR	0	1969–	Pu metal	..	4 800	Variable
France	Superphénix	2 900	1985–	Pu + U	31.5	..	1 600
India	FBTR	40	1985–	Pu + U	0.19	100	100
In commissioning							
Japan	Monju	714	1993–	Pu + U	5.9	1 100	540
Total Pu in first cores						**10 400**	
Planned							
Russia	BN800		before 2000	Pu + U	..	..	..
Japan	DFBR		after 2000	Pu + U	..	..	..
Europe	EFR		after 2000	Pu + U	..	..	..
India	PFBR		after 2000	Pu + U	..	..	..

[a] Only two full cores were fabricated and inserted; [b] The initial core was 12% enriched uranium.

Source: *Nuclear Engineering International, World Nuclear Industry Handbook 1992* (Reed International: London, 1991), Solonin, V. N., 'Utilization of nuclear materials released as the result of nuclear disarmament', paper presented at the international syposium on 'Conversion of Nuclear Warheads for Peaceful Purposes', Rome, 15–17 June 1992.

Table 7.2. Plutonium fuel fabrication facilities

Capacity is in tonnes of MOX per year; plutonium consumption is in kilograms fissile plutonium per year (kg Pu_{fiss}/yr).

Status/ Country	Facility	Operator[a]	Period of operation	Fuel	Capacity	Maximum plutonium consumption
Operating						
Belgium	Dessel DEMOX	BN	1973–	FBR/LWR	35	1 600
France	Cadarache ATPu	CEA	1970–89	FBR	15	4 000
France	Cadarache CFCa	Cogema	1990–	FBR / LWR	10 } / 15 }	2 200
FRG	Hanau BEW 1	Siemens	1972–92?	FBR/LWR	25–30	1 050
Japan	Tokai PFFF	PNC	1972–	FBR / ATR	1 } / 9 }	300
Japan	Tokai PFPF	PNC	1988–	FBR	5	900
UK	Sellafield	BNFL	1970–89	FBR	4	1 300
Planned						
Belgium	DEMOX P1	BN	Mid-1990s?	LWR	35	2 100
France	Marcoule Melox	Cogema	1996–	LWR	115	6 900
FRG	Hanau BEW 2	Siemens	1992–	LWR	80–120	7 200
Japan	Tokai PFPF	PNC	1993/4–	ATR	40	2 400
Japan	Rokkasho	JNFS	Late 1990s?	LWR	100?	6 000
Russia	Chelyabinsk	MAPI	1990s?	FBR	5?	300?
UK	Sellafield MDF	BNFL	1993–	LWR	8	400
UK	Sellafield SMP	BNFL	Late 1990s?	LWR	50–70	4 200

[a] BN = Belgonucléaire; CEA = Commissariat à l'Énergie Atomique; Cogema = Compagnie Générale des Matières Nucléaires; PNC = Power Reactor and Nuclear Fuel Development Corporation; BNFL = British Nuclear Fuels; JNFS = Japan Nuclear Fuel Services Company; MAPI = Ministry of Atomic Power and Industry.

mance and by balancing this against what is known about production at MOX fuel fabrication plants.

The United Kingdom

The two British fast reactors—the Demonstration Fast Reactor (DFR) and the Prototype Fast Reactor (PFR)—were located at Dounreay, the UKAEA site in the very north of Scotland. Dedicated fuel reprocessing facilities were also constructed there. Fuel fabrication was carried out at Windscale/Sellafield until 1989 when it was moved to Cadarache in France. Fuel-cycle strategy at British fast reactors has included reprocessing and plutonium recycle. The UKAEA and the CEA in France are unique in having fuel-cycle infrastructures in which fast-reactor fuel can be recycled more or less continuously. For instance, HEU fuel from the DFR was rapidly reprocessed (within six to eight weeks) so as to keep the UKAEA HEU inventory at a minimum during the 1960s and early 1970s.

The pressure on resources has been less intense where plutonium is concerned. A net surplus of civil plutonium existed in the UK when the PFR started operating in 1974, and for the first six years of operation discharged fuel was simply stored. Nevertheless, PFR fuel has been steadily reprocessed and recycled since 1980. The published amount of total plutonium in the fast-reactor fuel cycle in the UK—including that in reactor cores, in spent fuel held in storage ponds, in the process of extraction at the Dounreay reprocessing plant, being fabricated into MOX fuel, and in fresh fuel—is about 5 tonnes (to the nearest half tonne). This has changed little since figures were first published in 1982.[2] We would regard this as about the minimum required to keep the PFR reactor regularly fuelled, considering the delays and backlogs which develop within a fuel cycle (see chapter 6, section V). However, it is possible that some plutonium separated at Windscale/Sellafield before 1969, and thus belonging to the UKAEA, may also have been used in fabricating the PFR's first core.[3]

Although the PFR faces closure in 1994 with the withdrawal of British Government financing, the reactor may yet survive beyond that date. One of the principal reasons is the apparent donation of all the fuel produced for the first core of the German Kalkar SNR-300 to AEA Technology. After adaptation, this fuel may be inserted into the PFR.[4] This fuel originally contained some 1400 kg of fissile plutonium (see the discussion on Germany below). If the fuel can be modified and agreement reached on continued financing of PFR's research programme, this plutonium would presumably become the property of AEA Technology. It is not known whether this would reduce the amount of plutonium leased from the British utilities.

France

Both Phénix and Superphénix are fuelled with mixed plutonium/uranium oxide. However, only Phénix has a consistent record of refuellings. Its fuel cycle is spread between Marcoule (fuel reprocessing) and Cadarache (fuel fabrication).

Unfortunately there is little published data about fuelling strategy at the Phénix reactor, largely because it has been used as a source of weapon-grade plutonium (the extraction of plutonium from FBR blankets in France is discussed in chapter 3, section VII). It is only possible to draw certain upper bounds on how much plutonium it may have consumed. One way is to look at

---

[2] British Department of Energy, Press Release no. 178, 17 Oct. 1991, states that 5 t of plutonium has been sold by utilities to the UKAEA for fast-reactor R&D since 1969. This includes a small quantity of material used for other research purposes. *Parliamentary Debates, House of Commons [Hansard], Official Report*, 1 Apr. 1982, col. 169, states that 5.5 t of plutonium were sold or leased to the UKAEA between 1969 and 1981.

[3] Before 1969, the UKAEA owned all nuclear fuel in the UK. Ownership rights were transferred to utilities in 1969 and 1971 (see chapter 3).

[4] Ross, D., '£60 m reactor gift raises Dounreay hopes', *Glasgow Herald*, 9 Oct. 1991; Plummer, C., 'Fresh fuel begins to arrive at Dounreay', *John o'Groats Journal*, 15 Nov. 1991.

the reactor's operation, the other is to look at fuel-fabrication capacity. Phénix has a cumulative lifetime load-factor of about 56 per cent.[5] This implies that it has operated fairly consistently since it was commissioned. If we take a standard assumption that on average 40 per cent of its core was replaced each year,[6] and that reloads are enriched to 27 per cent with plutonium,[7] then over the 17 years of operation to 1990 some 7.9 tonnes of fissile plutonium could have been loaded into Phénix. Adding the core (830 kg Pu_{fiss}) and assuming one reload of fresh fuel had been fabricated (about 460 kg Pu_{fiss}) gives a total consumption of 9.2 tonnes of Pu_{fiss}. Subtracting the 6.7 tonnes of total plutonium (5 t Pu_{fiss}) which we estimate has been separated from Phénix and Rhapsodie spent fuel (see chapter 6, section V), and which we assume has been recycled, we can estimate that Phénix fuel has consumed about 4.2 tonnes of fissile plutonium (about 5.6 t Pu_{tot}). Considering that Phénix is similar in scale to the Dounreay PFR, and that more Phénix than PFR fuel has been reprocessed, this seems to be a reasonable figure. It also conforms with what is known about fast-reactor fuel fabrication capacity in France. At the time of writing Phénix was shut down. If it is brought back on-line, its annual consumption of plutonium would be about 400 kg Pu_{fiss}.

Superphénix, on the other hand, performed poorly from the moment it first operated at full power at the end of 1986 until it ceased operating for the last time in July 1990. By the end of 1991 it had a cumulative load factor of about 3 per cent. The reactor still contains its first core of fuel, although two reloads have been fabricated. In 1987, 13 fuel assemblies were removed from the reactor, but they were put back into the core when a fuel storage vessel began leaking in March of that year. Thereafter, the reactor has operated under a licence which allowed it to operate without a fuel store. In July 1992, Superphénix was effectively abandoned by the French Government for technical and political reasons.[8]

The Superphénix core plus two fabricated reloads contain about 8.6 tonnes of fissile plutonium (about 11.5 t of total plutonium).[9] Not all of this is French material. Électricité de France owns only 51 per cent of the reactor, the rest being divided between the Italian utility ENEL (33 per cent) and SBK, a consortium of other European utilities (16 per cent). The foreign component of the fuel is therefore about 4.2 tonnes. Taking Phénix and Superphénix together, it can be estimated that in 1992 about 8.6 tonnes of French fissile plutonium (4.2 t plus 4.4 t) and 4.2 tonnes of foreign-owned plutonium were dedicated to fast

[5] *Nuclear Engineering International, World Nuclear Industry Handbook 1992* (Reed Publishing: London, 1991), p. 20–21.

[6] This is the operating assumption at Superphénix. The fuel is designed to remain in-core for two operating cycles (28 months). See *Nuclear Engineering International* (note 5), p. 65.

[7] Phénix fuel core weighs 4.3 t (40% is 1.72 t). Fuel enrichments range from 19.2% to 27.2% fissile plutonium. NERSA, *The Creys-Malville Plant*, 1987, p. 8.

[8] MacLachlan, A., 'French Premier steps in personally to Superphénix debate', *Nucleonics Week*, 2 July 1987; MacLachlan, A., 'Creys-Malville staff have begun the final search for the leak', *Nucleonics Week*, 13 Aug. 1987; MacLachlan, A., 'Prime Minister orders more work, public inquiry for SuperPhenix', *Nucleonics Week*, 2 July 1992.

[9] First core contains 4.8 t of Pu_{fiss}, two reloads would contain 3.8 t of Pu_{fiss}.

reactors in France (about 11.5 t of French total plutonium and about 5.6 t of foreign-owned total plutonium).

Japan

Plutonium has been used in six different R&D applications in Japan: the Japan Atomic Energy Research Institute's fast critical-fuel assembly; PNC's deuterium critical assembly; the Joyo fast reactor; the Fugen Advanced Thermal Reactor (ATR); the Monju fast reactor; and MOX assemblies in operating LWRs. The two critical assemblies are experimental rigs together containing some 950 kg of total plutonium, and do not require refuelling.

The Joyo reactor is an experimental facility which has been fuelled with plutonium and uranium. No fuelling information has been published for the reactor, but it is known that the initial core contained 12 per cent enriched uranium. MOX reloads were fabricated at the Tokai Plutonium Fuel Fabrication Facility (PFFF), starting in 1972/73. Since October 1988 Joyo fuel has been produced at the new Plutonium Fuel Production Facility (PFPF) plant.[10] By 1989 some 1600 kg of total plutonium had been used in producing this fuel. Annual reloads at Joyo reportedly consume 60–120 kg of plutonium. Assuming that the first reload took place in 1978, a maximum of 1440 kg could have been loaded into the reactor.[11] Assuming that the Plutonium Fuel Development Facility (PFDF) plant was dedicated to producing the first core for the Monju reactor since then, it can also be assumed that no more Joyo fuel had been produced by the end of 1990.

The 165-MWe Fugen ATR, which began operating in 1978, has also been fuelled with plutonium. Fuel for the reactor has been produced since 1978 at the ATR line of Tokai's PFFF plant. This has capacity for producing 9 tonnes of fuel containing about 170 kg of total plutonium per year. However, plutonium is not essential to the operation of the reactor—it can be substituted by low-enriched uranium. It is therefore difficult to make firm plutonium consumption estimates based on assumptions about regular fuelling with the material. Based on information provided by Japanese industry and government officials, Fugen plutonium consumption by 1989 was 1250 kg.[12] Given that annual reloads at Fugen contain between 70 and 170 kg of plutonium, it can be estimated that by the end of 1990, 1400–1600 kg of total plutonium had been consumed in ATR fuel.

[10] Nakano, H., Kaneki, H., Shishido, T. and Suzuki, Y., 'Automation improves safety at Japan's Tokai MOX plant', *Nuclear Engineering International*, Dec. 1989, pp. 27–29. The PFDF began operating in 1965. 1.5 t of MOX fuel were produced for irradiation test studies. The PFFF began operating in 1972 and produced fuel for Fugen and Joyo. The FBR line at the PFPF began operating in 1988 on Joyo and Monju fuel. An ATR line is due to be added to PFPF to produce fuel for the planned Ohma DATR.

[11] Plutonium enrichments range from 20% to 40%; Berkhout, F., Suzuki, T. and Walker, W., 'Surplus plutonium in Japan and Europe: an avoidable predicament', MIT Working Paper, MITJP 90-10, Cambridge, Mass., Sep. 1990, p. 9.

[12] Berkhout *et al.* (note 11), p. 9. The Fugen fuel core weighs 34.3 t.

A new prototype fast reactor, Monju (280 MWe), is due to begin operation during 1993. The initial core will contain about 920 kg of fissile plutonium (about 1200 kg Pu_{tot}).[13] Fuel production began at the PFPF plant in 1990. Because of problems with fuel pellet fabrication in late 1991, it is expected that the first core will not be delivered to the reactor until January 1993.[14] Thereafter, annual reloads of fuel are expected to contain about 540 kg Pu_{fiss} (720 kg of total plutonium).[15]

If it is assumed that by the end of 1990 about one-third of the Monju initial core of fuel had been fabricated, then Japanese critical assemblies and fast-reactor fuel had absorbed between 4.4 and 4.6 tonnes of total plutonium by then.

Two new plutonium-fuelled demonstration reactors may be constructed in Japan over the next decade: a Demonstration ATR at Ohma (606 MWe, scheduled operation March 2001)[16] and a Demonstration FBR for which a site has not yet been chosen. If plutonium is the chosen fuel for the DATR its initial core will contain about 2400 kg fissile plutonium (3100 kg Pu_{tot}).[17] Fuel for the reactor is due to be produced at a new ATR line at the PFPF facility at Tokai, with an annual capacity of 40 tonnes of MOX fuel. Annual reloads (19 t of fuel) are planned to have a fissile enrichment of 3.1 per cent. About 780 kg of total plutonium would therefore be required annually.

The size and fuelling requirements of the DFBR have not yet been published. In the longer term there are also plans for a follow-up reactor to the DFBR, although no details have yet been given about the size or timing of this reactor.[18] However, assuming reactors in the 750–1000 MWe range, they would each require some 3 to 4 tonnes of fissile plutonium for their initial cores, with annual reloads of 1.2 to 1.6 tonnes Pu_{fiss}. As yet, Japan has no firm plans to build a fast-reactor fuel reprocessing plant.

Germany

Plutonium has been used in the production of fuel for two German fast reactors, although only one of these has operated.[19] The KNKII research reactor

[13] Akebi, M., et al., 'Building Monju—Japan's prototype FBR', Nuclear Engineering International, Oct. 1991, pp. 37–44. Fuel assembly weight: 0.0298 t, 108 inner core assemblies (15% enriched), 90 outer-core assemblies (20% enrichment). Reload enrichments will be 16% and 21% respectively.

[14] Usui, N., 'PNC claims it is catching up on Monju breeder fuel delay', Nucleonics Week, 2 Jan. 1992, p. 13; Usui, N., 'Fuel fabrication, backfit delay start-up of Monju fast breeder', Nucleonics Week, 5 Mar. 1992.

[15] Together Joyo and Monju are expected to require about 600 kg Pu_{fiss} (about 800 kg Pu_{tot}) per year. Japan Atomic Energy Commission, Nuclear Fuel Recycling in Japan, Report by the Advisory Committee on Nuclear Fuel Recycling (provisional English translation), Tokyo, Aug. 1991, pp. 13–14.

[16] The planned commissioning date for the DATR has been put back twice in two years: 'Ohma ATR project postponed by one year', Atoms in Japan, Dec. 1990, p. 24; and 'Demonstration ATR project postponed for one year', Atoms in Japan, Dec. 1991, p. 22.

[17] Nuclear Engineering International (note 5), p. 88. Initial core: 95 t (enrichment 2.5%); reload enrichments 3.1%.

[18] Japan Atomic Energy Commission, 1991 (note 15), pp. 13–14.

[19] During the 1960s, Euratom bought small quantities of plutonium for experimental use at the Masurca (France) and SNEAK (FR Germany) reactors. Under the first agreement, 350 kg of plutonium for critical

(17 MWe) at Karlsruhe operated with a fast core between 1977 and 1991. Two separate fuel cores were loaded at the reactor, the first inserted in 1977 and discharged in 1982, the second in 1982 and discharged in 1991.[20] The first core was composed mostly of HEU, although it also contained 110 kg of US-origin plutonium (see chapter 6, section V).[21] Assuming the second core contained plutonium fuel, and further assuming a mean enrichment of 59 per cent,[22] the total consumption of plutonium by the reactor was 430 kg of fissile plutonium (about 550 kg Pu_{tot}). A third core was fabricated, but never loaded. Assuming that this was equivalent to the second, the plutonium total consumption in KNK II fuel was 1.2 tonnes Pu_{tot}.

Information about the first core of the ill-fated Kalkar (SNR-300, 295 MWe) reactor is much better. Although fully constructed, and provided with a complete first core of fuel, this reactor did not operate before the project was abandoned in 1991.[23] The fuel assemblies for the initial core were fabricated at the Belgonucléaire facility at Dessel (40 per cent) and the Siemens plant at Hanau (60 per cent).[24] They contained 1400 kg of fissile plutonium (1750–1800 kg total plutonium), giving an enrichment of about 27 per cent.[25]

The majority of these assemblies are still stored at Dessel and Hanau. There have recently been moves to use the fuel at the Dounreay PFR in Scotland (see the section on the United Kingdom, above). It is assumed here that the plutonium still belongs to the owner of the Kalkar reactor.

In sum, about 3.25 tonnes of total plutonium have been used in fuel assemblies for two German fast reactors. Since neither reactor will operate again, no further fast reactor demand for plutonium is anticipated.

Russia and Kazakhstan

Information about fast reactors in the former Soviet Union is still sparse. Nominal capacities and core sizes are known, but nothing is published about their operating performance, or the fuelling strategy adopted.

Two fast reactors, one sited at Byeloyarsk in Russia (BN600, 560 MWe) and the other at Chevchenko in Kazakhstan (BN350, 135 MWe), are known to have

assemblies was sold to Euratom. A second agreement made provision for an extension of this amount to 500 kg. However, this was not taken up, the total amount purchased coming to 460 kg. About one-third of this material was used in Germany. Personal communication from W. Stoll, 25 Mar. 1992.

[20] *Nucleonics Week*, 5 Sep. 1991.

[21] 'KNK II—An experimental power station equipped with a fast core', *Nuclear Engineering International*, Jan. 1979, pp. 41–44.

[22] *Nuclear Engineering International* (note 5), p. 77.

[23] Hibbs, M., 'BMFT confirms scuttling of SNR-300, Siemens to take Interatom in hand', *Nucleonics Week*, 28 Mar. 1991.

[24] Hibbs, M., 'Bonn will retain Pu custody but may transfer inventory', *Nucleonics Week*, 30 Apr. 1990.

[25] This plutonium came from a variety of sources. About 550 kg originated from LWR fuel reprocessing of RWE fuel at La Hague, about 500 kg was plutonium from French Magnox reactors, about 170 kg came from the Netherlands reactor, Borselle, and most of the remainder was separated at WAK. Because of the different isotopic compositions of this material, it is difficult to estimate precisely what this fissile quantity is equivalent to in total plutonium.

used plutonium fuel, but only on an experimental basis.[26] They have been largely fuelled with HEU. This was justified by the large surplus stocks of HEU held in the Soviet Union, and continuing technical difficulties with plutonium fuel fabrication.[27]

Large-scale plutonium use was not envisaged until the South Urals fast reactor project at Chelyabinsk was announced in 1986 (originally four reactors, scaled back to a single unit). To supply this project, construction of a new fuel fabrication plant (Tsech 300) began at Chelyabinsk in 1984. However, neither the reactor nor the fuel factory have been completed, and there are doubts about their future. Solonin suggests that little plutonium (probably between 600 kg and 700 kg) has been consumed in fast-reactor fuel in Russia and Kazakhstan.[28]

If the South Urals 1 reactor (BN800) were completed, then plutonium consumption would increase markedly. The reactor is officially projected to be brought on-line in 1997. Assuming a design output of about 750 MWe, and assuming a core in proportion to output, the reactor should have a core weighing about 20 tonnes, containing about 3 tonnes fissile plutonium. Reloads would be about 1.2 tonnes Pu_{fiss} per year.

The United States

There are no electricity-producing fast reactors currently operating in the USA. Three small research reactors—EBR1, EBR2 and Fermi 1—date from the early 1950s and the early 1970s, but none was fuelled with plutonium.

Fast-reactor research has continued at a number of sites under the auspices of the US Department of Energy, however. Two facilities, the Fast Flux Test Facility (FFTF) at Hanford, and the Zero Power Plutonium Reactor (ZPPR) at the Idaho National Engineering Laboratory (INEL), have been the main consumers of plutonium outside the weapon programme. Much of the plutonium contained in them is believed to have come from the UK (see chapter 3).

The FFTF reactor (400 MWth) has been operating since 1980. Four core loadings of fuel (7.5 tonnes of MOX) containing just over 2.9 tonnes of fuel-grade plutonium (nominal 12 per cent ^{240}Pu) were fabricated for the reactor by the Kerr-McGee Corporation during the 1970s.[29] Although there were plans to produce additional fuel at Los Alamos during the late 1980s, no further reloads have been produced since then, and it is doubtful whether the reactor will oper-

[26] Kazanskii, Y. A., et al., 'Physical characteristics of the BN-600 reactor', Soviet Atomic Energy, vol. 55, no.1 (July 1983), pp. 441–47; and Kazachkovskii, O. D., et al., 'Development and experience of operating fast reactors in the Soviet Union', Soviet Atomic Energy, vol. 54, no. 98–95 (Apr. 1983), pp. 270–79; Solonin, V. N., 'Utilization of nuclear materials released as the result of nuclear disarmament', paper presented at the International Symposium on Conversion of Nuclear Weapons for Peaceful Purposes, STES, Rome, 15–17 June 1992.

[27] A pilot fuel fabrication facility has been build and apparently operated successfully.

[28] Solonin (note 26).

[29] Cochran, T. B., Arkin, W. M., Norris, R. S. and Hoenig, M. M., Nuclear Weapons Databook, Vol. II: US Nuclear Warhead Production (Ballinger: Washington, DC, 1987), p. 76.

ate for much longer. The ZPPR Project contains 3.8 tonnes of plutonium, 3.4 tonnes fuel-grade, 0.2 tonne weapon-grade and 0.2 tonne reactor-grade.[30]

Total plutonium consumption in US fast-reactor research has therefore been about 6.7 tonnes.

Summary of plutonium use in fast reactors: past and projected

Taking all these figures together we can estimate that fast reactors and critical assemblies have so far accounted for between 37 and 38 tonnes of plutonium. Summary past consumption figures, together with estimates for projected demand from operating and planned reactors are shown in table 7.3.

In drawing up the future scenarios in table 7.3, the following assumptions have been made:

1. The high scenario: the British PFR operates until 2000; Phénix is brought back on-line in 1992 and operates until 2005; all the Japanese reactors are constructed and operate as planned (Joyo, Fugen, Monju, DATR and DFBR); and BN800 in Russia is commissioned in 1997.

2. The low scenario: the PFR is closed in 1994; Phénix is shut down permanently in 1993; Joyo and Fugen operate until 2000 and Monju operates from 1993 at 50 per cent capacity in Japan; BN350 and BN600 continue to be fuelled with enriched uranium, while BN800 is not completed.

As with the projections for plutonium separation, the results show a very wide range of possibilities. However, one trend is clear. The historical figures show that fast reactors and Japan's ATR consumed about one-quarter of the plutonium separated from power-reactor fuel to the end of 1990 (122 t Pu_{tot}). In the next two decades, the proportion of separated plutonium loaded at fast reactors will fall. Because of stalled programmes elsewhere, future fast reactor consumption of plutonium will be highly dependent on new reactors being built in Japan and Russia.

Given the technical, economic and political problems facing fast reactors in recent years, it seems unlikely that the expansive Russian and Japanese policies represented in the high scenario will be realized. As a result, plutonium consumption in fast reactors (and ATRs) may be closer to the low than the high scenario presented here. In our view, 5–10 tonnes of plutonium is a more plausible range for consumption in these R&D reactors in both decades.

As R&D reactors can only consume a small proportion of plutonium arisings under any of the above scenarios, the pressure to find another way of disposing of plutonium will continue to grow if reprocessing goes ahead as planned. Recycling in thermal reactors is the only substantial alternative.

---

[30] Cochran *et al.* (note 29), p. 76.

Table 7.3. Plutonium in fast-reactor fuel cycles, end of 1990, and in 1991–2000 and 2001–10

Figures are in tonnes of total plutonium.

Country	End of 1990	1991–2000[a]		2001–10[a]	
		High	Low	High	Low
France	17.1	3.2	–	2	–
FRG	3.25	..	..	..	..
Japan	4.4–4.6	8	5	30–35	4
Russia/Kazakhstan	0.6–0.7	7	..	16	–
UK	5	..	..	..	..
USA	6.7	..	..	..	..
Total	37.05–37.35	18.2	5	48–53	4

[a] The projected plutonium consumption figures do not take material recycling into account. Assuming that some reprocessing of fast reactor fuel would continue if the reactors were kept operating, and that separated plutonium were recycled, the figures above are over-estimates of the actual amount burned up.

V. Plutonium use in thermal reactors

Interest in recycling plutonium in thermal reactors has a considerable history. During the 1950s there was a presumption in Europe and the USA that reprocessing capacities would be greater than those required to feed fast reactors, at least for a transitional period. In the interim, excess plutonium would be used to fuel thermal reactors. The result was a co-operative research programme between Euratom and the US Atomic Energy Commission.

This work continued through the 1960s and early 1970s during which time MOX elements were loaded at LWRs in 10 countries.[31] For the most part, small amounts of plutonium were involved. The most significant research programmes were in Belgium, centred around work at the BR3 PWR, and in Germany where between 1966 and 1977 about 214 MOX fuel assemblies were loaded at five LWRs.[32]

During the 1970s doubts increased about future plutonium supply. Difficulties were experienced in both Europe and the USA in bringing large reprocessing plants into operation, and by 1977 President Carter had put forward his policy aimed at restricting plutonium production and use. In a period when a belief in developing commercial fast reactors still existed, this restriction on plutonium supply meant that interest in thermal recycle faded in most countries, even in Belgium and Germany.

---

[31] Bairiot, H., 'Laying the foundations for plutonium recycle in light water reactors', *Nuclear Engineering International*, Jan. 1984, pp. 27–33.

[32] Schlosser, G. J. and Winnik, S., 'Thermal recycling of plutonium and uranium in the Federal Republic of Germany: strategy and current status', IAEA-SM-294/33, in *The Back-end of the Nuclear Fuel Cycle: Strategies and Options* (IAEA: Vienna, 1987), pp. 541–49.

During the 1980s, MOX recycling returned as an option in Western Europe and Japan as fast reactor prospects faded and surpluses of separated plutonium became apparent. MOX received greater priority first in Germany, and a little later in Switzerland, France and Belgium. Towards the end of the decade, big plans for thermal MOX recycle were also revealed in Japan.[33] By early 1992 there were also signs that Russia was seriously considering thermal recycle as a way of using both its civil and military stocks of plutonium, even if there are serious doubts over whether such plans could be implemented.

The MOX production process

The MOX fuel fabrication process is more capital-intensive than uranium fuel production because additional safety precautions are necessary when handling plutonium, especially when it is extracted from high burn-up fuels. Therefore the key to keeping MOX prices down is to maintain high throughputs at fabrication plants. The plants are also sensitive to the quality of plutonium that is available. Current facilities cannot handle plutonium containing more than 1.3 to 1.5 per cent of americium-241 (a decay product of ^{241}Pu). For typical LWR plutonium this means that once it has gone through the final purification stage in reprocessing, it cannot be stored for more than two to three years. New plants due to begin operating in the 1990s will be able to handle plutonium containing up to about 2.5 per cent ^{241}Am, thereby extending storage times by a further two or three years, making standard burn-up and fuel cooling-time assumptions.

For these reasons, plutonium fuel fabrication has considerably higher costs than the fabrication of enriched-uranium fuel. A study by the OECD's Nuclear Energy Agency (NEA) found that MOX fabrication prices are between three and six times higher than the prices for uranium fuel fabrication.[34] The hope has been that production costs will in time fall with increased volume of output. However, future cost reductions through economies of scale will be partly off-set by higher capital-intensities due to the need to protect operators handling plutonium from high burn-up fuels.

To maintain high throughputs and rapid fuel fabrication, the rates of reprocessing and MOX fuel fabrication should be closely matched. One of the problems is that fabrication is likely to lag considerably behind reprocessing, despite MOX programmes being brought forward in a number of countries to ensure increased consumption. Reprocessing schedules are generally set by the reprocessors, who seek to maximize fuel throughput at their plants. Unless these schedules are adjusted downwards, substantial stocks of plutonium could accumulate at reprocessing sites. These stocks would quite rapidly become unusable because of the buildup of ^{241}Am. It remains to be seen whether reprocessors will be prepared to reduce the rate of production to avoid this happen-

---

[33] 'Japan's fuel recycling policy: AEC formulates plan to use plutonium mainly for LWRs', *Atoms in Japan*, Aug. 1991, pp. 4–10.
[34] OECD/NEA, *Plutonium Fuel: An Assessment* (OECD: Paris, 1989), pp. 68–73.

ing. The dilemma they face is that by slowing the rate of reprocessing, they will increase their costs.

Once committed to reprocessing, the utility is therefore left with little room for manœuvre. Plutonium production is decided by the reprocessor, and the rate at which it is fabricated into fuel is determined to a large extent by existing capacities at fuel factories. For individual utility companies it is therefore very difficult to match supply with demand.

In normal circumstances, orders for MOX fuel must be placed by a utility with the fabricator some 18 months before plutonium is dispatched to the fabrication plant. On arrival, the material is typically stored for up to 10 months, followed by a chemical analysis. Fabrication may take about two months, with a further two-month delay before the fuel is delivered to the reactor. In all, the time between initial order and delivery, assuming all goes to plan, is between 30 and 36 months.

VI. National programmes for thermal plutonium recycle

No single account exists of thermal plutonium recycle activities around the world. Since plutonium fuel use has not been concentrated in one particular reactor type, and because it has until recently been on a pre-industrial scale, tracing historic plutonium consumption must be based almost entirely on data about amounts of fuel fabricated, and their plutonium enrichments.

Germany

The first German MOX R&D programme ran from 1966 until 1977. It was funded by the federal government and was concentrated at three reactors: the VAK reactor at Kahl; the Gundremmingen A BWR; and the Obrigheim PWR. Altogether some 21 tonnes of MOX fuel were produced containing about 600 kg of fissile plutonium.

A second 'pilot' MOX agreement was concluded between Siemens and German nuclear utilities in 1980. This covered the period 1982–88 and involved the loading of 201 fuel assemblies (84 t MOX, about 2.3 t fissile plutonium) at seven PWRs and two BWRs.[35] As with the first programme, all of this fuel was manufactured at the Alkem plant at Hanau. Figure 7.1 shows the production record at Hanau between 1966 and 1990.

---

[35] Participating reactors were: Obrigheim, Neckarwestheim 1, Grafenrheinfeld, Unterweser, Grohnde, Brokdorf, Phillipsburg 3 (PWRs), and Gundremmingen and Kruemmel (BWRs). Schmiedel, P., 'Experience with plutonium recycling in the Federal Republic of Germany', paper to the Uranium Institute Thirteenth Annual Symposium, 7–9 Sep. 1988, London; Schlosser and Winnik (note 32); Hibbs, M., 'German utilities bracing for MOX fuel cost increases', *Nuclear Fuel*, 6 Jan. 1992, pp. 10–11; Brandsetter, A., Schmiedel, P. and Stoll, W., 'Einsatz von Plutonium in LWR und SBR', *Atomwirtschaft*, Aug./Sep. 1984, pp. 453–58; Schmiedel, P., 'Recycling and its implications for uranium demand: the German perspective', paper presented to the Uranium Institute Fifteenth Annual Symposium, London, 5–7 Sep. 1990. This paper states that by Sep. 1990 Siemens had manufactured almost 70 000 fuel rods weighing 133 t and containing 3.7 t of fissile plutonium (a mean plutonium enrichment of 2.78%).

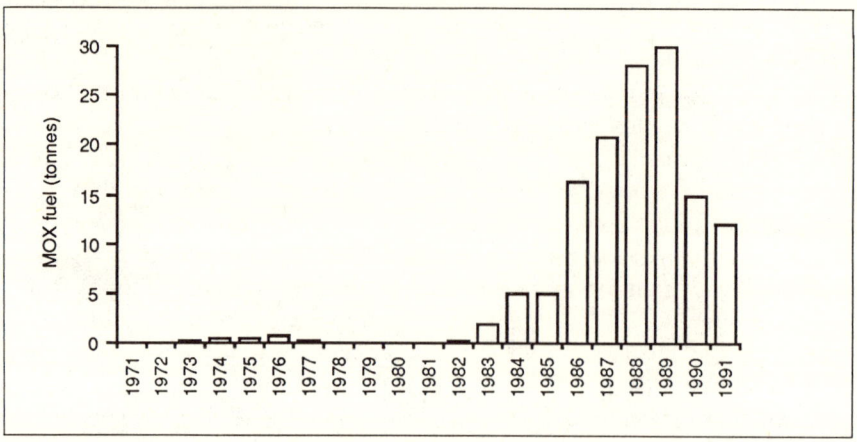

Figure 7.1. LWR–MOX production at the German Hanau fuel fabrication plant, 1972–91

Sources: Brandsetter, A., Schmiedel, P. and Stoll, W., 'Einsatz von Plutonium in LWR und SBR', *Atomwirtschaft*, Aug./Sep. 1984, pp. 453–58; and Schmiedel, P., 'Recycling and its implications for uranium demand: the German perspective', paper to the Uranium Institute Fifteenth Annual Symposium, London, 5–7 Sep. 1990.

On 1 January 1989 a new MOX agreement between Siemens and the utilities came into force. The primary motivation for the utilities was to have to hand a way of dealing with the large quantities of plutonium which were due to be returned to them during the 1990s. An increase in fabrication capacity was therefore planned at Hanau, and construction began in late 1987 on a new bunkered facility. Capacity could then be raised from 35 to about 120 tonnes MOX per year by the mid-1990s. According to the 1989 agreement, German utilities would take between 70 and 80 per cent of the fuel produced at Hanau up until 1998.

In parallel, applications were made for licences to load MOX at all but three (making a total of 18) of the German commercial LWRs for which MOX licences had not been issued. At the time of writing, 10 reactors held plutonium fuel licences, but difficulties were being encountered negotiating the remainder. Licences now exist for the loading of up to 172 assemblies each year, containing between 3.5 and 4 tonnes of fissile plutonium (4.7–5.3 t total plutonium).[36] For operational reasons, however, the actual number of assemblies that could be loaded under existing licences ranges from about 60 in the early 1990s to about 110 by 2000 (containing from 1.2 to 2.5 t of total plutonium respectively).[37]

---

[36] Dibbert, H. J., 'Strategien des Brennstoffkreislauf', *Atomwirtschaft*, Feb. 1991, pp. 83–88.

[37] Vereinigung Deutscher Elektrizitätswerke (VDEW), *Strategieüberlegungen zur Brennelemententsorgung und Verwertung von Plutonium und wiederaufarbeitetem Uran: Gegenwärtige Situation und langfristige Perspektiven*, Frankfurt, Sep. 1989, table 10.

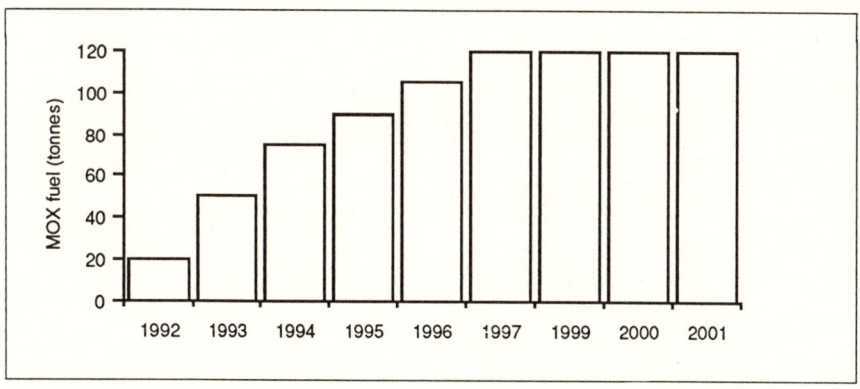

Figure 7.2. Projected LWR–MOX production at the German Hanau BEW2 fabrication facility, 1992–2001

Source: Vereinigung Deutscher Elektrizitätswerke (VDEW), *Strategieüberlegungen zur Brennelemententsorgung und Verwertung von Plutonium und wiederaufarbeitetem Uran: Gegenwärtige Situation und langfristige Perspektiven*, Frankfurt, Sep. 1989, table 10.

The outlook remains very uncertain. There have been delays in the completion of the new fabrication facility; licences granted to the facility by an outgoing state government in 1991 are being challenged in the courts (the state of Hesse, in which Hanau is located, is now ruled by a Social Democratic Party (SPD)–Green coalition government); limits had already been placed on production capacity at the original Hanau facility when in June 1991 a plutonium leak led to its complete closure;[38] and concern has been voiced by German utilities over the prices being charged for MOX fuel.[39]

Figure 7.2 shows the anticipated production of MOX fuel at Hanau, assuming that the new facility operates as planned, and that this fuel can be loaded at reactors. The reactor bottle-neck is perhaps more crucial than the restraints on MOX production. If all the separated plutonium due to be returned to Germany during the 1990s is to be used, then the additional eight reactor MOX-loading licences are essential. In view of state government opposition to plutonium use in critical states such as Lower Saxony, and the economic penalties associated with recycle, it must be considered unlikely that the current plan will come to fruition. Perhaps the most optimistic forecast is that reactors which now have licences will be routinely fuelled with MOX, with an annual fissile plutonium consumption rising during the 1990s from under 1 tonne to a little over 2 tonnes.

---

[38] 'Incidents close Hanau plant', *Nuclear Engineering International*, Aug. 1991, p. 3.
[39] Hibbs (note 35).

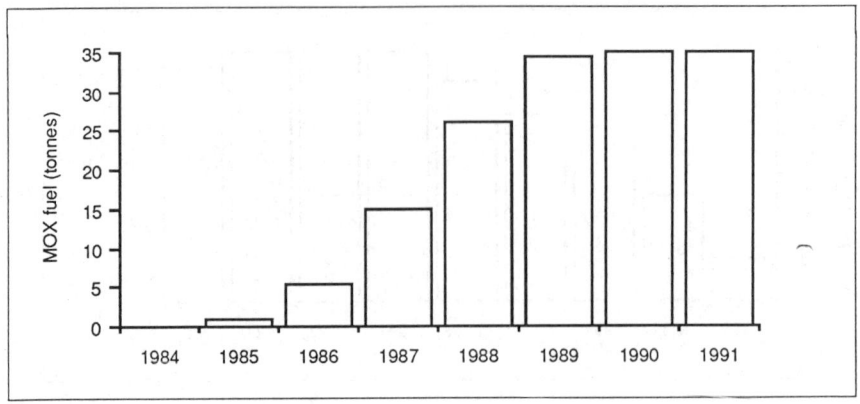

Figure 7.3. LWR–MOX production at the Belgian Dessel fuel fabrication plant, 1984–91

Source: Dr H. Bairiot, Belgonucléaire, Private communication with the authors, 24 Aug. 1990.

Belgium

Belgonucléaire, together with Siemens, has been a pioneer of LWR–MOX fuel fabrication, beginning with the pilot-scale production at Mol of plutonium fuel for the BR3 reactor in 1962/63. New production capacity, capable of fabricating FBR and LWR–MOX fuel, was established at Dessel in 1973.[40] Production remained at a low level until 1983/84, with some 240 fuel assemblies being produced for two BWRs (Garigliano and Dodewaard) and one PWR (Chooz), besides BR3 fuel.[41] During this period somewhat under 10 tonnes of MOX were manufactured, containing about 550 kg total plutonium.[42]

In 1984, under the joint COMMOX initiative between Cogema and Belgonucléaire, the plant (renamed DEMOX) was backfitted to give it a rated capacity of 35 tonnes of MOX per year. It was gradually brought to full capacity in 1989 (see figure 7.3). Over 80 per cent of this capacity has been taken up with orders for French reactors, the rest being produced for BR3, Swiss and German reactors. By the end of 1990, 117 tonnes of fuel containing 5.73 tonnes of total plutonium had been produced in this new phase of production. Total plutonium consumption in LWR–MOX production at Dessel until the end of 1990 was therefore about 6.3 tonnes.

COMMOX holds contracts with the French utility Électricité de France (EdF) to supply about 25 tonnes of MOX annually up to 1994. These contracts,

---

[40] Le Bastard, G., 'MOX fuel fabrication: present and future', paper given to the Uranium Institute Fourteenth Annual Symposium, London, 6–8 Sep. 1989.

[41] A total of some 220 BR3 assemblies containing about 300 kg total plutonium were produced at Dessel. See van Dievoet, J., Bairiot, H., Le Bastard, G., Pakarek, H. and Roepenack, H., 'MOX fuel and its fabrication in Europe', paper given to the 7th Pacific Basin Nuclear Conference, San Diego, 4–8 Mar. 1990; *Nuclear Engineering International* (note 5), p. 85.

[42] Bairiot, H., Deramaix, P., Mostin, N., Trauwer, E., Vanderbrock, Y., 'Foundations for the definition of MOX fuel quality requirements', paper presented at the Characterization and Quality Control of Nuclear Fuel Conference, Karlsruhe, 19–21 June 1990.

together with further demand from Swiss, Belgian and German utilities, appear to assure production at Dessel until the mid-1990s. Belgium's own annual consumption of MOX fuel is expected to rise from 3 tonnes in 1991 to 7.8 tonnes in 1996–2000.[43] As in Germany, the objective is to recycle in thermal reactors all the plutonium separated in reprocessing. It seems that 35 tonnes per year is achievable, given the demonstrated capacity and the lack of legal obstacles for BN when compared with Siemens at Hanau.

The company also hopes to build a second module at Dessel (P1) which would double capacity to 70 tonnes MOX per year by 1995/96. At that rate, between 3 and 3.5 tonnes of total plutonium could be processed into fuel.

France

Plutonium recycling in LWRs was not considered by EdF until the early 1980s when, because of reduced fast reactor forecasts, a series of studies was commissioned on its technical and economic characteristics.[44] This led in 1985 to the signing of agreements between EdF and French fuel-cycle companies which envisaged the construction of about 90 tonnes of MOX fabrication capacity in France by the mid-1990s. This was subsequently increased, and the Melox plant, under construction by COMMOX at Marcoule, now has a design capacity of 115 tonnes per annum (5.6 to 6.1 tonnes of total plutonium). Before Melox comes on stream, EdF is depending on MOX production at Dessel, although some of its needs will also be met by the CEA's plutonium fuel factory at Cadarache which was backfitted to produce LWR–MOX in 1989. By 1990 production at Cadarache had reached 15 tonnes MOX (670–710 kg total plutonium) per year.

MOX was first loaded into a French reactor in September 1987, when 16 assemblies were inserted into the St Laurent B1 reactor.[45] Since then MOX has been loaded at three further 900-MWe class PWRs (St Laurent B2 and Gravelines 3 and 4). By the end of 1990, 37 tonnes of MOX (containing about 1.75 t total plutonium) had been loaded at French reactors, most of it produced at Dessel.[46] EdF plans to fuel 16 of its reactors (all in the 900-MWe class) with one-third core of MOX when the Melox plant begins operating in 1995/96.[47] Anticipated plutonium consumption at Melox is shown in figure 7.4.

Even under the most optimistic MOX fuelling scenario, EdF still expects to have a surplus of between 19 and 25 tonnes of Pu_{fiss} by the turn of the century

[43] Berkhout, F. and Walker, W., 'Spent fuel and plutonium policies in Western Europe: the non-nuclear weapon states', *Energy Policy*, July/Aug. 1991, pp. 553–66.
[44] EdF, Service des Combustibles 'Retraitement recyclage', Paris, 6 Mar. 1990, mimeo.
[45] 'Saint Laurent inaugure les "MOX"', *Revue Generale Nucleaire*, no.5 (Sep./Oct. 1987), p. 490.
[46] Simon, M. A., 'Recycling developments in France', paper presented at the Uranium Institute Fourteenth Annual Symposium, London, 5–7 Sep. 1990,.
[47] Rome, M., Salvatores, M., Mondot, J. and Le Bars, M., 'Plutonium reload experience in French Pressurized Water Reactors', *Nuclear Technology*, vol. 94 (Apr. 1991), pp. 87–91.

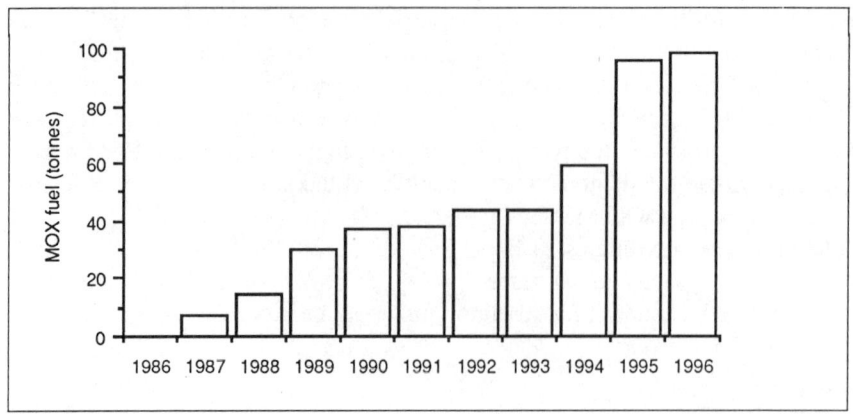

Figure 7.4. Past and projected MOX fuel reloads at Électricité de France reactors, 1987–96

Source: Simon, M. A., 'Recycling developments in France', paper presented at the Uranium Institute Fourteenth Annual Symposium, London, 5–7 Sep. 1990.

(taking account of the demise of Superphénix) unless the planned rate of reprocessing is reduced.[48] EdF is, however, unique among utilities in having retained in its contracts substantial influence over the rate at which its fuel is reprocessed. If there were hitches in the MOX recycling programme, it would be able to slow down the supply of plutonium.

Switzerland

Details of thermal recycle in Switzerland are scarce. No official federal government policy exists on the use of plutonium,[49] and the utilities themselves have published little about their activities. As they are not members of fast reactor consortia, their only use for separated plutonium is as LWR–MOX. The first demonstration loading of MOX assemblies took place at the Beznau reactor in 1978.[50]

In pursuing a recycle policy since the early 1980s, the utilities have been constrained by US nuclear export policy. Under the US Nuclear Non-Proliferation Act, 're-transfers' (as from Cogema to a MOX fuel fabricator) of plutonium produced from US-origin uranium must receive approval from the US Congress and regulatory bodies. As Switzerland is not a member of the European Community, its nuclear relations with the USA do not come under the US–Euratom agreement which covers fuel transfers to other West European countries. Because of concerns in the USA about Swiss nuclear export policies,

[48] EdF, 'Combustible MOX: aspects techniques, economiques et strategiques', Service des Combustibles, Paris, Nov. 1989, mimeo, table 2.
[49] OECD/NEA (note 34), Annex F, p. 114.
[50] van Dievoet, F. *et al.* (note 41).

the US Congress has insisted on approvals being made case-by-case, which has been time-consuming for the Swiss to arrange.

Nevertheless, four approvals were made during the 1980s—for 200kg in 1985; 108 kg in 1988; and 150 kg and 132 kg in 1990,[51] making a total of 590 kg fissile plutonium (about 780 kg of total plutonium). It is also known that 612 kg of total plutonium were fabricated into MOX fuel at Dessel during 1990 for insertion into the Beznau reactors. Previously transferred plutonium had been manufactured into MOX at Hanau for the Goesgen reactor. Assuming that the last two US approvals covered material transfers to Dessel and that these were used to produce fuel in 1990, a little over 1 tonne of total plutonium (about 20 t of MOX fuel) has been returned to Switzerland in the form of MOX.

Utility policy is to recycle plutonium separated under existing reprocessing contracts, but not to extend these contracts.

Japan

As in France, the policy of recycling plutonium in LWRs was slow to develop in Japan. Until the mid-1980s the government and utilities still believed that plutonium demand from fast reactors and ATRs would be sufficient to absorb the material which would be returned by European reprocessors during the 1990s.

Although a small-scale demonstration programme was inaugurated with the insertion of two fuel assemblies at the Tsuruga 1 BWR in 1986,[52] it was not until the Ministry of International Trade and Industry (MITI) published its 1986 Nuclear Energy Vision that plutonium recycle became policy.[53] The plan is that two 800-MWe reactors (one BWR and one PWR) will be fuelled with one-quarter core MOX by the mid-1990s. By the late-1990s four 1000-MWe LWRs are due to be loaded with one-third core MOX, this number rising to 12 reactors by 2005. Annual plutonium demand for the 12-reactor plan would be 4 tonnes of fissile plutonium per year.[54] However, if utilities have difficulty gaining local political consent for recycling plutonium in the full complement of 12 reactors, the rate of consumption could be considerably lower. In the longer term, Japanese utilities are aiming to build their own MOX fabrication plant at

---

[51] The last transfer was from La Hague to Dessel. See Knapik, M., 'US appears set to bless commercial use of plutonium in Swiss nuclear plants', *Nuclear Fuel*, 3 Sep. 1990.

[52] This was followed in 1988 with four assemblies (containing just 15 kg of total plutonium) being loaded at the Mihama 1 PWR—Matsuoka, Y. and Abeta, S., 'Mihama MOX trials meet with success', *Nuclear Engineering International*, Dec. 1989, pp. 24–25. If the Tsuruga assemblies contain a similar percentage of plutonium then the total plutonium loaded as MOX in Japan is about 20 kg.

[53] MITI, *Nuclear Energy Vision*, Tokyo, Sep. 1986, pp. 39–42.

[54] Samejima, K., 'The use of plutonium in Japan', paper presented to the Uranium Institute: Fourteenth Annual Symposium, London, 5–7 Sep. 1990; and AEC (note 15). Samejima estimates that MOX plutonium consumption will increase from 140 kg Pu_{fiss} in 1992 to 3960 kg Pu_{fiss} in 2000. At this rate, some 50% of the 32 t Pu_{fiss} expected to be separated from Japanese fuel in Europe would be manufactured into LWR–MOX fuel.

Table 7.4. Plutonium consumption in LWR–MOX fuel fabrication, up to the end of 1990

Figures for MOX fuel are in tonnes; figures for plutonium are in kilograms; figures in brackets assume that total plutonium contains 75 per cent fissile material.

Country	MOX fuel	Fissile plutonium (Pu_{fiss})	Total plutonium (Pu_{tot})
Germany	145	3 800	(5 100)
Belgium	127	(4 600)	6 300
France	17	800	(1 100)
Japan	< 1	(15)	20
Total	**290**	**9 200**	**12 500**

Rokkasho-mura, with a capacity of around 100 tonnes of MOX fuel per year. Until then, they are planning to meet their MOX requirements by having the fuel fabricated in Europe.

One possibility that has been discussed by Japanese utilities and European MOX fabricators is that most of Japan's plutonium arisings in France and Britain could be fabricated into MOX fuel before its return to Japan. Transporting plutonium as MOX could ease the political difficulties of returning it, and plutonium could be stored in Europe longer if held in MOX fuel (americium buildup would be less of a problem). Assuming that the fabrication would be carried out in Belgium, France or at the planned facility at Sellafield, and that some 35 tonnes of plutonium were involved, Japanese demand could occupy around 30–40 per cent of European MOX fabrication capacity throughout the 1990s.[55] The difficulty that Japanese utilities face, however, is that they may be constrained in placing MOX contracts until the MOX demonstration programme in Japan is well advanced. A clear pattern is therefore not likely until the late 1990s.

VII. Summary of plutonium use in thermal reactors: past and projected

The estimates of past consumption made above are summarized in table 7.4. About 12.5 tonnes of total plutonium had been consumed up to the end of 1990 in the production of about 290 tonnes of MOX fuel.

As shown above, future production of LWR–MOX fuel is prone to great uncertainty. While several new plants are planned, it is not yet clear whether some of these will actually be built (P1, SMP at Sellafield, and the plant at Rokkasho), or how those that are being built (BEW 2 at Hanau and Melox) will operate. Even estimating production at existing plants such as BEW 1 at Hanau is difficult because of the political situation in Germany.

---

[55] This assumes plutonium enrichment in MOX fuel of 5%, and a total Belgian, French and British MOX fabrication capacity of 265 t per year.

Table 7.5. Estimated LWR–MOX fuel outputs in 1991–2000 and 2001–10, high and low scenarios

Figures are in tonnes.

Country/ facility	1991–2000 High	1991–2000 Low[a]	2001–10 High	2001–10 Low[a]
Belgium				
DEMOX	350	260	350	260
DEMOX P1[b]	140	–	350	–
France				
CFCa	150	100	–	–
Melox[c]	450	300	1150	750
FRG				
BEW 1	35	12	–	–
BEW 2	840	–	1 200	–
UK				
MDF[d]	60	–	–	–
SMP[e]	215	–	700	–
Total	**2 240**	**672**	**3 750**	**1 010**

[a] The low scenarios for operating plants assume 60% load factors.
[b] Assumes Dessel 2 begins operation in 1996 and reaches full capacity in 1998.
[c] Assumes that Melox begins operation in 1996 and reaches full capacity in 1999.
[d] Assumes MDF comes on stream in 1993, and reaches full capacity in 1994.
[e] Assumes SMP comes into operation in 1997 and takes three years to reach full capacity.

Estimates made here have also been based on fabrication capacity, not on assumptions about whether the fuel produced will actually be loaded at reactors. This may be a source of overestimation since fuel will not be fabricated until MOX transport and loading licences have been approved. This adds a new layer of complexity which may throw up unexpected results. Nor have we attempted to track the movement of MOX fuel assemblies to reactors, even where they cross international boundaries (German/Swiss and Belgian/French). In future, especially with the prospect of Japanese utilities contracting for MOX fabrication in Europe, these cross-boundary transfers are likely to become more frequent and large-scale.

The fuel fabrication capacities assumed in making plutonium consumption forecasts below are set down in table 7.5. These two views of the future by no means exhaust the possibilities. The low scenarios we propose are heavily weighted towards the Melox plant in France. If there is to be a rationalization of European MOX capacity, it is likely to lead to greater concentration, not to more plants operating at lower capacities. The Melox plant, assuming it operates to plan, seems the most likely to survive in such circumstances. The high scenario assumes that all plants consistently operate at maximum capa-

Table 7.6. Projected plutonium consumption by MOX loaded in LWRs in 1991–2000 and 2001–10, high and low scenarios[a]

Figures are in tonnes.

Country/ facility	1991–2000		2001–10	
	High	Low	High	Low
Belgium				
Dessel 1	19	14	25	18
Dessel 2	9	–	25	–
France				
CFCa	8	6	–	–
Melox	27	18	80	53
FRG				
BEW 1	2	0.5	–	–
BEW 2	48	–	84	–
UK				
MDF	3	2	–	–
SMP	13	–	49	–
Total	129	40.5	263	71

[a] Plutonium enrichment in MOX fuel is assumed to reach 5% in 1995; 6% in 1996–2000; and 7% from 2001 onwards. This conforms with assumptions made in Dievoet, J., Bairiot, H., Le Bastard, G., Pakarek, H. and Roepenack, H., 'MOX fuel and its fabrication in Europe', paper presented at the 7th Pacific Basin Nuclear Conference, San Diego, Calif., 4–8 Mar. 1990.

cities. We have not taken any account of a possible fabrication plant at Rokkasho-mura, because no firm plan has yet been set out. The fuel-fabrication figures in table 7.5 are expressed in terms of plutonium consumption in table 7.6.

If MOX production follows the 'high' scenario, thermal recycle will undergo a dramatic growth during the 1990s, rising from consuming 13 tonnes of plutonium in the previous 20 years to using 10 times that amount in this decade. As shown, there are strong doubts that this can be attained. There is currently no clear commercial incentive to recycle plutonium, except as a way of avoiding punitive plutonium storage costs and the political embarrassment of accumulating large stocks of this sensitive material. The 'high' scenario is unlikely, but it is hard to predict how far below this level MOX production will fall.

VIII. Commercial and R&D plutonium use compared with quantities separated

The market for plutonium is now largely supply- rather than demand-driven. Heavy investments have been made in reprocessing capacity, and utilities in Europe and Japan are at present contractually bound to use it. But plutonium has not become a competitive fuel, or the fast reactor a competitive technology, contrary to the nuclear industry's expectations in the 1970s when these commitments were first made.

Rather than reduce supply to match falling demand, attempts are being made at a rather late hour to raise demand to absorb a fixed supply. Demand for plutonium is being 'constructed'. The outcome will depend on more than progress in building adequate industrial capacities to supply MOX fuels, and in securing funds to meet the extra costs involved. Political obstacles also have to be overcome, including the problems associated with transporting the plutonium.

Future trends in plutonium consumption are therefore extremely unpredictable. Moreover, the prospects vary widely across the five countries which are due to have to deal with the largest quantities of plutonium. These countries can be divided into three main groups:

1. *Britain and Russia.* Neither country has experience of recycling plutonium in its thermal reactors. The UK has no plans to do so in future, and Russia is constrained by economic and safety concerns. Civil plutonium arisings in these countries are therefore likely to remain surplus to requirements.

2. *France.* Has a strong institutional commitment to recycling, and the political authority to push it through. However, its fast-reactor programme is in difficulty, leaving a gap in demand which even an aggressive thermal recycling programme is unlikely to fill completely.[56]

3. *Germany and Japan.* While Japan remains committed to expanding its plutonium economy, German enthusiasm has waned. There are serious political obstacles to large-scale recycling in both countries, particularly regarding transportation and reactor licensing in Japan, and the licensing of MOX fabrication capacity in Germany.

According to the estimates drawn up in chapter 6, present reprocessing plans will give rise to 310 tonnes of plutonium by the end of 2000, and 545 tonnes by the end of 2010. By the end of 1990, 50 tonnes had been recycled in fast and thermal reactors, and 72 tonnes were as yet unused. In table 7.7 we compare these arisings with two scenarios of plutonium consumption.

The *maximum scenario* is the sum of the 'highs' in tables 7.3 and 7.6. It assumes that the remaining fast reactor plans are realized; that all thermal MOX fabrication capacity is constructed according to schedule, allowing large thermal reactor recycling programmes to get under way; and that plutonium enrichment levels in thermal MOX fuel rise towards 7 per cent early in the next century.

Instead of the low scenarios set out in tables 7.3 and 7.6, a *moderate scenario* is presented here which envisages that recycling will continue in both fast and thermal reactors, but at a slower pace. It also assumes that political obstacles in Germany and Japan cause plutonium recycling in those countries to be more

---

[56] Even with an active MOX programme, EdF still expects to have a surplus of plutonium by the turn of the century of up to 50% of the estimated 53 t Pu_{fiss} which is due to be separated from the utility's fuel by 2000. See EdF (note 48).

Table 7.7. Cumulative supply and consumption of total civil plutonium, ends of 1990, 2000 and 2010

Figures are in tonnes.

	End of 1990	End of 2000	End of 2010
Present reprocessing plans			
Plutonium separation	122	310	545
Maximum recycling scenario			
Plutonium consumption (FBR)	37	55	105
Plutonium consumption (LWR)	13	142	404
Balance of supply and demand (current plans)	+ 72	+ 113	+ 36
Moderate recycling scenario			
Plutonium consumption (FBR)	37	45	53
Plutonium consumption (LWR)	13	93	227
Balance of supply and demand (current plans)	+ 72	+ 172	+ 265
Balance of supply and demand (stretch-out scenario)[a]	+ 72	+ 106	+ 90

[a] See table 6.12. The stretch-out scenario envisages production of 122 tonnes in 1991–2000, and 126 tonnes in 2001–10.

inhibited than in France and Belgium. This scenario rests on the following assumptions: 8 tonnes of plutonium (the mid-point of the 5–10 tonne range suggested in section IV above), will be consumed in each decade in fast reactors;[57] French and Belgian MOX fabrication plants will be operated at 80 per cent, and the German plant at 50 per cent, of planned output; Japanese production of thermal MOX fuel after 2000 will match German levels; and plutonium enrichment levels will average 5 per cent across the whole period.

Under the high-production and -use scenario, plutonium consumption is raised slightly above production after 2000, gradually leading to a draw-down of surplus stocks. A surplus of 113 tonnes in 2000 is reduced to just 36 tonnes in 2010. In the moderate-use scenario, the rate of consumption never approaches the rate of production. This scenario envisages the world plutonium surplus growing rapidly to 172 tonnes in 2000, and increasing further to 265 tonnes in 2010. The balance of supply and demand achieved under the moderate plutonium recycling scenario, with a lower rate of reprocessing considered under the 'stretch-out scenario' presented in chapter 6, shows consumption nearer in line with production, but with the surplus still ranging from 90 to 106 tonnes in the period up to 2010.

It is up to the reader to decide which of the above scenarios is the more plausible, but our view is that the high scenario is very unlikely to come about, and that even the moderate-use scenario could turn out to be optimistic. We

[57] This corresponds roughly to the fuelling of Phénix in France, and Joyo, Fugen and Monju in Japan with no further fast reactors or ATRs being commissioned before 2010.

therefore conclude that very large surpluses of civil plutonium will develop unless the rate of reprocessing is substantially reduced. These figures also suggest that there is almost no room for the absorption in currently planned MOX recycle programmes of large amounts of plutonium released from nuclear weapon dismantling.

8. Civil highly enriched uranium inventories

I. Introduction

Almost 200 research, test and power reactors world-wide use HEU fuels. Almost all of this HEU was produced in military enrichment plants in the nuclear weapon states (NWS). Much of the spent HEU fuel from non-nuclear weapon states (NNWS) has been returned to the NWS for reprocessing. In the USA, the recovered HEU was kept by the USA and recycled in US military production reactors at the Savannah River site in South Carolina. With the end of the cold war, this practice has stopped.

In the 1970s, relatively large inventories of civilian HEU, mostly containing over 90 per cent ^{235}U, and thus weapon-grade material, were accumulating in Europe and elsewhere, requiring large numbers of international shipments. Although all this material outside the NWS was under International Atomic Energy Agency (IAEA) or Euratom inspection, there was concern that international commerce in HEU would increase the chance that some of it would be diverted and used to make nuclear weapons.

There has also been concern that a NNWS might seize its stock of safeguarded HEU and put it to military purposes. This possibility received widespread attention as a result of the Persian Gulf War. Iraq had in its possession about 26 kg of essentially unirradiated 80 and 90 per cent enriched uranium from the former USSR and France. The USA and its allies worried that Iraq might divert this HEU for use in an atomic bomb.

Because of these concerns, the USA and other Western countries have been co-operating since the late 1970s to reduce the amount of civilian HEU in international commerce. The main focus of these efforts has been the development of new low-enriched uranium (LEU) fuels for thermal research reactors.

II. Civil HEU suppliers

By far the largest supplier of civil HEU has been the USA, which over the years has exported about 25 000 kg of HEU, containing about 18 000 kg of ^{235}U.[1] It has also supplied almost 30 000 kg of weapon-grade uranium to domestic thermal, research and power reactors.[2]

---

[1] Albright, D., 'Civilian inventories of plutonium and highly enriched uranium', eds P. Leventhal and Y. Alexander, *Preventing Nuclear Terrorism* (Lexington Books: Lexington, Mass., 1987), pp. 265–91; Letter from Armando Travelli, Reduced Enrichment for Research and Test Reactors (RERTR) Program Manager, to Leonard Weiss, Staff Director, US Senate Committee on Governmental Affairs, 9 June 1989.

[2] Albright (note 1); Cochran, T., Arkin, W., Norris, R. and Hoenig, M. M., *Nuclear Weapons Databook, Vol. II: US Nuclear Warhead Production* (Ballinger: Cambridge, Mass., 1987), appendix D.

The average amount of HEU exported by the USA has declined dramatically during the past decade. The amount of ^{235}U in HEU exported during 1970–77 averaged 660 kg per year.[3] This amount decreased to an average of 380 kg per year during 1978–82, and decreased further to an average of 160 kg per year during 1983–88.

The other major supplier of civilian HEU has been the USSR, although it mainly supplied HEU to its own research and fast reactors. We estimate that about 10 000 kg of HEU went to Soviet research reactors, and almost 17 000 kg was exported to Eastern Europe, Iraq, Libya, North Korea and Viet Nam. A larger quantity of HEU has been supplied to Soviet fast breeder reactors.

The UK, China and France have supplied small amounts of HEU fuel to a few countries. France was to have supplied Iraq with roughly 30–40 kg of weapon-grade uranium a year to fuel the 40 MW Osirak research reactor. After the Israeli bombing of this reactor, this contract was stopped, but not before about 12 kg of weapon-grade material had been delivered.

III. Reactors using HEU fuels, 1992

Over half of the research reactors that use HEU fuel have lifetime cores, and require no additional HEU fuel. The others require 1200–1300 kg of ^{235}U in HEU fuels per year. We estimate that over half of this material is used in US and former Soviet research reactors (see appendix D).

Two fast breeder reactors in the former USSR, with a combined electrical generating capacity of about 950 MWe, use HEU fuels (see chapter 7). Each year, they are estimated to require roughly 1000–1500 kg of ^{235}U in HEU (containing mostly 21 per cent ^{235}U). Two high-temperature power reactors in the USA and Germany would have required about 300 kg of weapon-grade uranium per year, but they have recently been shut down. Plans to build more of them have been shelved. As a result, HEU now plays no part in Western nuclear power production programmes.

In addition to the amount of HEU used as fuel each year, a larger amount of HEU is in the 'pipeline'. This includes all HEU that is held by fuel fabricators, stored at reactor sites as fresh or irradiated fuel, reprocessed or transported between any of these facilities. If the time for the HEU fuel to go through the entire cycle from enrichment to reprocessing is four years, roughly four times the amount required by reactors each year would be in circulation.

IV. Converting to low-enriched uranium fuels

Because of concerns about the diversion of HEU fuel, the USA started the Reduced Enrichment for Research and Test Reactors (RERTR) programme in 1978 to eliminate the need for HEU fuels in most civilian thermal research reactors. Since then, this programme has worked with countries throughout the

[3] Travelli to Weiss (note 1).

world to develop LEU fuels that can substitute for HEU fuels. The reduction in annual US HEU exports can be traced in part to the efforts of this programme.

The RERTR programme currently targets about 40 non-US research reactors with a power of at least 1 MWe that have received HEU from the USA or other Western countries.[4] These reactors together require about 450 kg of ^{235}U in HEU per year. There are another 40 or so reactors that have been supplied by Western countries that use HEU fuels, but these are small and probably have enough HEU fuel at their sites to last until they are shut down.

Nine of the targeted reactors which together used to require HEU containing 76 kg of ^{235}U per year, have converted to LEU fuels. Another 29 of them, which annually require HEU with a total of 275 kg of ^{235}U, are in various stages of conversion. Final conversion of these reactors could occur by the mid-1990s. In addition, five US university reactors, which are not considered target reactors of the RERTR programme, have also converted.

Overall, the RERTR programme is expected to reduce by roughly one-third the annual amount of ^{235}U in HEU required by thermal research reactors. Because of a lack of funding in the USA, however, the RERTR programme is not developing new low-enriched fuels for the three largest targeted research reactors that the USA supplies (one in Belgium and two in France), which together require about 100 kg of ^{235}U per year. It has been proposed that the programme include a fourth large reactor in the Netherlands, but the Netherlands Government has refused to participate. In addition, the US Government has refused to convert its own research reactors that use HEU fuels. Together these US reactors require about 400 kg of ^{235}U per year. Chinese research reactors have also not been targeted by the RERTR programme. Furthermore, breeder research and power reactors that use HEU fuel have never been singled out for conversion to LEU fuels.

In mid-1992 the US and Russian governments began discussions on a joint programme to convert research reactors in the former USSR, including facilities in Russian nuclear weapon laboratories. According to a Department of Energy official, the funds for this effort would come from the USA.

V. Reprocessing of HEU fuels

Because irradiated HEU fuel contains a considerable amount of residual ^{235}U, it is typically reprocessed and the HEU recovered. Belgium, France, the UK, the USA and probably the former USSR have all reprocessed civil HEU fuels.

France and the UK have reprocessed HEU fuel from thermal and fast research reactors. The UK has recovered about 1.7 tonnes of HEU from thermal research reactor fuel, although the source of this information did not state the amount of ^{235}U contained in the recovered material.[5]

---

[4] The rest of this section is based on Travelli, A., 'The RERTR program: a status report', Paper presented at the 1991 International Meeting on Reduced Enrichment for Research and Test Reactors, Jakarta, 4–7 Nov. 1991.

[5] 'The Dounreay fuel cycle facilities of AEA Fuel Services', *Atom*, Jan. 1992, pp. 18–20.

The USA has reprocessed most of its spent domestic research reactor fuel. This includes HEU fuel from experimental breeder reactors. It has also reprocessed foreign spent HEU fuel when it was the supplier of the HEU. Up to the end of 1988 the USA had received about 3800 kg of ^{235}U in foreign spent HEU fuel.[6] Spent fuel from the high-temperature reactors has not been been reprocessed, and is currently stored pending final disposal.

In 1988 the USA decided not to take back more foreign spent HEU fuel. The former USSR is not believed to have taken back foreign spent research reactor fuel. It is unclear whether either country will take back foreign spent fuel in the future. As a result, countries with large research reactors are having to store their fuels, and the UK and France are considering offering expanded reprocessing services to foreign customers.[7] Unlike the reprocessing services provided by the USA, these would require the contracting country to take back the recovered HEU and the reprocessing wastes.

VI. Civil HEU inventories

It is possible to make a crude estimate of the amount of HEU that remains in civil programmes. Most of the civil HEU used in US and perhaps former Soviet civil research reactors has been reprocessed and returned to the military programmes. In the military programmes, however, it forms a small fraction of the total amount of recovered HEU.

The total amount of HEU remaining in countries supplied by the USA can be estimated. From 1978 to the end of 1988, the USA issued export licences for 3289 kg of ^{235}U in HEU, and accepted the return of 1629 kg of ^{235}U.[8] Another 1570 kg of ^{235}U were consumed in reactors during this period. During the 1980s, US exports of ^{235}U therefore nearly matched the return of ^{235}U in spent fuel and the amount consumed in reactors. Therefore, HEU inventories in countries supplied by the USA have not increased significantly during the past decade.

For earlier periods, however, more was exported than returned. Only about 2150 kg of ^{235}U in spent HEU fuel had been returned by the end of 1977, out of a total of about 14 700 kg exported by then. Assuming that the returned HEU matched that returned to the USA in the 1980s (in which case about half the ^{235}U was consumed), then about 10 400 kg of ^{235}U remains in Western countries supplied by the USA.

---

[6] Cochran et al.(note 2); US Department of Energy official, private communication with the authors, Feb. 1992.

[7] Marshall, P. and MacLachlan, A., 'DOE inaction leads AEA Technology, Cogema to weigh reprocessing MTR fuel', *Nuclear Fuel*, 17 Feb. 1992.

[8] Letter from Armando Travelli, RERTR Program Manager, to Leonard Weiss, Staff Director, US Senate Committee on Governmental Affairs, 19 Apr. 1988; Travelli (note 6).

Table 8.1. Estimated cumulative inventories of domestic and exported HEU used in civil thermal research and test reactors, by enricher country, end of 1990

Figure are in tonnes.[a]

Enricher country	Supply ^{235}U	HEU	Returned to supplier for reprocessing or storage ^{235}U	HEU	Estimated inventory ^{235}U	HEU
Former USSR						
Domestic[b]	9.0	10.0	4.7	6.2	1.0	1.1
Exported[c]	1.0	1.7	–	–	0.7	1.4
USA						
Domestic[d]	27.0	29.0	14.0	18.0	1.3	1.6
Exported[e]	18.0	25.0	3.8	5.4	6.2	13.0
China[f]						
Domestic and exported	0.9	1.0	0.30	0.40	0.3	0.4
France						
Domestic and exported	< 0.5	< 0.5	< 0.5	< 0.5	–	–
UK						
Domestic and exported	< 0.5	< 0.5	< 0.5	< 0.5	–	–
Total	**57.0**	**68.0**	**24.0**	**31.0**	**9.5**	**17.0**

[a] Figures exclude fast reactors, except those outside the USA using US-supplied HEU.

[b] For research reactors in the former USSR, the amount of HEU estimated as returned to the supplier is the total supply of irradiated fuel after accounting for the amount of HEU estimated to be at the fuel fabricators and the reactors. This latter value is estimated to be equivalent to four times the annual requirement (300 kg ^{235}U per year; see appendix D), minus the amount consumed by two years of irradiation. This assumes that each reactor has assigned to it about two-years' worth of fresh fuel and two-years' worth of irradiated fuel. It is also assumed that 40% of the ^{235}U in irradiated fuel is consumed, either through fission or conversion into ^{236}U.

[c] See note b. In this case the total annual requirement is about 60 kg ^{235}U per year. The spent fuel was apparently not returned to the USSR.

[d] See note b. In this case a 50% consumption rate and an annual requirement of 435 kg ^{235}U per year is assumed. Excluded from these assumptions is 2.8 t of ^{235}U in 3.0 t of Project Rover fuel that was only slightly irradiated.

[e] See text. It is assumed that the returned HEU fuel was originally 90% enriched.

[f] See note b. In this case the annual requirement is 100 kg ^{235}U per year. A 40% consumption rate is assumed.

Source: Appendix D, see notes 1 and 2; and IAEA, *Nuclear Research Reactors in the World* (IAEA: Vienna, 1991).

A significant proportion of this 10 400 kg of ^{235}U was either consumed in reactors, remains in their cores, or was recovered in Europe. A fraction—an estimated 2000 kg—remains in storage or in the fuel cycle as fresh material.[9] Assuming that about half of the remainder has been consumed, at least 6200 kg

[9] Much of this material is in the form of unused fuel for the German high-temperature reactor and the KNK-II, both of which are shut down.

of ^{235}U (4200 kg in irradiated form and 2000 kg in fresh form) in HEU remains in these US-supplied research reactor programmes. If we include 6200 kg of ^{238}U in the original fresh HEU fuel (initial HEU enrichment of 70 per cent on average) and any ^{236}U produced in the fuel, the amount of HEU supplied by the USA and not returned is about 13 tonnes.

We have made similar estimates for HEU inventories elsewhere. The results are summarized in table 8.1. The total inventory of HEU dedicated to civil thermal reactors (estimated enrichment of 55 per cent) is about 17 tonnes.

The IAEA stated in its 1990 Annual Report that it safeguarded 11.3 tonnes of HEU located in NNWS that are parties to the NPT, and 0.4 tonnes in countries outside the NPT (see chapter 12). Euratom similarly stated that it safeguarded around 13 tonnes at the end of 1990 (the amount being greater as it includes relatively larger amounts of material in France and the UK than in Eastern Europe, Canada and Japan). The IAEA safeguards no HEU in the NWS. These provide aggregate quantities for a certain group of countries. What is needed is a country-by-country break-down of civil HEU inventories. Until this becomes available, the picture will remain incomplete.

While our estimates for material under safeguards are roughly equivalent to the quantities declared by the IAEA and Euratom, our estimated total is higher, mainly because it includes unsafeguarded HEU in all the NWS.

Part IV
Material inventories and production capabilities in the threshold states

The appreciation by the threshold countries that nuclear weapons confer enormous advantages upon the country having them and can affect the imbalance in manpower, natural resources, industrial, potential and military strength, is a very big incentive for nuclear proliferation and has contributed a lot to the desire of these countries to keep all options open. (A.Q. Khan, father of Pakistan's enrichment programme, in Khan, A. Q., 'The spread of nuclear weapons: militarization or development', Sadruddin Aga Khan (ed.), *Nuclear War, Nuclear Proliferation, and their Consequences* (Clarendon Press: Oxford, 1986), pp. 423–24.)

Any country that wants to keep open its nuclear weapon options must have a supply of separated plutonium or highly enriched uranium. Although the inventories of these materials in threshold countries are small compared to those found in the acknowledged nuclear weapon states or in the major industrialized countries, they pose a disproportionate risk because only kilogram quantities of these materials are needed for each nuclear weapon. Even a single nuclear weapon in the hands of certain national leaders could be destabilizing, increasing regional tensions and the likelihood of nuclear war.

Part IV is organized into three chapters that discuss the fissile material production capabilities and inventories of three major groups of countries. Chapter 9 discusses the *de facto* nuclear weapon states, India, Pakistan and Israel. These nations are not parties to the NPT, and have accumulated unsafeguarded inventories of fissile material. Although all three countries deny that they have nuclear weapons, a wide variety of intelligence and scientific sources reveal that they either have fully assembled, deliverable nuclear weapons or could assemble them within a short period of time.

Chapter 10 evaluates the nations that are known to have secret fissile material production programmes or ambitions, but that have not accumulated significant inventories of unsafeguarded fissile materials. This group includes Iraq, Iran, North Korea and Algeria. All of these countries, except Algeria, have signed the NPT, but are viewed as potential 'false adherents' to the Treaty. Although its programme was curtailed by the Persian Gulf War, Iraq might reconstitute its programme if it is allowed to do so. Iraq's ability to develop a large weapon programme in secret, despite having signed the NPT, demonstrates that knowledge of the scope and size of nuclear programmes, and the whereabouts of nuclear materials, is vital to international security.

Chapter 11 examines countries that have recently taken steps away from nuclear weapons. The chapter first considers Argentina, Brazil and South Africa, who have recently committed not to build nuclear explosives and to allow international inspections of their fissile material production programmes. The chapter concludes with a discussion of Taiwan and the steps it has taken to bolster confidence that it will not pursue nuclear weapons.

In these three chapters, information about each country's fissile material production capabilities acts as the basis for estimating inventories of plutonium and highly enriched uranium. Because this information is in many cases a closely guarded secret, several of these estimates are highly uncertain. However, over the past several years a great deal of information has become available that makes the estimates more dependable.

9. *De facto* nuclear weapon states: Israel, India and Pakistan

Israel is widely believed to have deployed nuclear weapons. India and Pakistan have the capability to deploy weapons quickly. Because of the secrecy surrounding nuclear programmes in these countries, little definitive, public information exists about the number of weapons they might possess, or the size of their current or planned stockpiles of separated plutonium and highly enriched uranium.

Nevertheless, during recent years several research groups, members of the media and other observers have published information about these nuclear programmes. We have expanded upon these earlier assessments to provide an estimate of the stockpiles of fissile materials in these countries.

I. Israel

Israel operates several highly secret nuclear facilities at a site near Dimona that are aimed at producing plutonium for nuclear weapons. This site contains a small production reactor and a plutonium separation facility.

Although plutonium appears to be Israel's main route to nuclear weapons, Israel has pursued uranium enrichment, at least as a research activity. Israel has also been suspected of having diverted about 100 kg of weapon-grade uranium from a US nuclear facility during the 1960s, although Seymour Hersh has recently presented evidence that the diversion did not occur.[1]

Plutonium production at Dimona

The reactor

Israel's main source of plutonium is a French-supplied reactor. Early reports said that it generated 24 MWth when it began operating in December 1963. The reactor's power, however, appears to be substantially higher, giving Israel a significantly larger plutonium production capability.

The lack of conclusive information about the power levels reached by the Dimona reactor creates the largest uncertainty in estimating Israel's plutonium inventory. The only detailed account of the construction and initial size of the Dimona reactor was published in 1982 by Pierre Pean.[2]

[1] Hersh, S. M., *The Samson Option* (Random House: New York, 1991).
[2] Pean, P., *Les deux bombes* (Fayard: Paris, 1982). Translated into English by the US Congressional Research Service (CRS).

According to Pean, the initial Franco-Israeli agreement in 1956 called for France to supply a nuclear reactor of the 'EL-3' type, an 18-MWth research reactor that was moderated by heavy water with cooling provided both by the heavy water and by air.[3] However, according to Pean, when the French team charged with building the plutonium extraction facility read the reactor designs, they were surprised that the reactor power was two to three times larger than indicated in the original Franco-Israeli agreement.[4] They also found that the cooling ducts were three times larger than those needed for a 24-MW reactor. With these upgrades in the original design, the head of the construction team and his deputy concluded that the Dimona reactor would be capable of producing as much plutonium as the G1 reactor at Marcoule, France. This is a 38-MWth gas-cooled graphite-moderated reactor that is capable of producing about 13 kg of weapon-grade plutonium per year if operated all year at full power.[5]

Public upper-bound estimates of the reactor's power range from about 70 to 150 MW. The highest power rating is derived from details about the plutonium separation plant at the Dimona complex made public in 1986 by Mordechai Vanunu, a former nuclear technician at the Dimona separation facility. According to a US official, however, who spoke on condition his name would not be used, the consensus of people in the government who have studied the information supplied by Vanunu is that the reactor's power has not reached 150 MW. He said that the power of the Dimona reactor has probably never exceeded 70 MW. This latter estimate of the power is consistent with Pean's information about the size of the cooling ducts.

Spector and Smith report that US Government specialists believe that the reactor operated initially at 40 MW, and might have been enlarged to 70 MW prior to 1977, when Vanunu began working at Dimona.[6] Assuming that they are correct, the reactor could produce about 8.8 to 15 kg of weapon-grade plutonium a year, if it operated between 40 and 70 MWth, 60 per cent of the time on average.[7]

The separation plant

Vanunu's revelations have provided the public with its only detailed glimpse into Israel's plutonium separation programme.[8] He described an underground

[3] See Pean (note 2). In CRS translation: 'Suez or the "Musketeers" of the Bomb', p. CRS-4. Details of the EL 3 reactor can be found in International Atomic Energy Agency, *Directory of Nuclear Reactors, Vol. 2: Research, Test, and Experimental Reactors* (IAEA: Vienna, 1959), pp. 295–300.

[4] See Pean (note 2). In CRS translation: chapter VI, 'EL 102', p. CRS-11.

[5] See Pean (note 2). In CRS translation, p. CRS-12; Turner, S. E., *et al.*, 'Criticality studies of graphite-moderated production reactors', SSA-125, Southern Sciences Applications, Jan. 1980, table 2.1 (report prepared for the Arms Control and Disarmament Agency). The G1 reactor produces about 0.92 g of plutonium per MWth-day.

[6] Spector, L. S. and Smith, J. R., *Nuclear Ambitions* (Westview Press: Boulder, Colo., 1990), p. 160.

[7] These estimates assume that 1 g of weapon-grade plutonium is produced per MWth-day of heat output.

[8] 'Revealed: the secrets of Israel's nuclear arsenal', *Sunday Times*, 5 Oct. 1986; and Barnaby, F., *The Invisible Bomb* (I.B. Taurus: London, 1989).

facility at Dimona, supplied by the French, that included floors dedicated to separating plutonium from irradiated fuel by the PUREX process, converting the plutonium into metal, and shaping the metal into bomb components.

Vanunu also supplied quantitative information about the flow of plutonium through two sections of this plant. He said that:

1. The plutonium plant produces an average of nine metallic buttons a week, each of which contains about 130 grams of plutonium. Over an eight-month production cycle, the plant would process a total of about 40 kg of plutonium.

2. The standard flow rate of dissolved uranium and plutonium through one section of the plant is 20.9 litres per hour. The concentration of uranium is 450 grams per litre, and the plutonium concentration is 170–180 milligrams per litre (or 0.39 mg of plutonium per gram of uranium). This corresponds to a production of 22–23 kg of plutonium, assuming that the plant operated continuously for eight months a year. Vanunu, however, stated that the flow rate normally exceeded the standard rate by 150–175 per cent, implying an annual production of 33–40 kg of plutonium per year.[9]

That the separation plant was achieving outputs in the above range is supported by a knowledgeable French nuclear source, who did not dispute most of Vanunu's claims in the *Sunday Times*. He said, however, that a yearly capacity of 40 kg a year might be 'a little on the high side'.[10]

Statements that the power of the reactor reached 150 MW result from calculations based on a steady-state output of 40 kg of plutonium per year.[11] An alternative explanation is that the years when Vanunu worked at the plant were a time when a backlog of irradiated fuel was processed, resulting in a temporary increase in the plant's annual throughput. Evidence for this hypothesis is that Vanunu was hired as part of a large group of technicians in 1977, and laid off in 1985 with about 180 other workers. This latter explanation, if true, would imply that the output of this reprocessing plant is not a reliable indicator of the plutonium production rate of the Dimona reactor.

Plutonium inventory

The above discussion illustrates the difficulty of determining Israel's plutonium inventory. As a result, we have developed various estimates based on hypothetical power histories of the reactor, labelled A–E in table 9.1.

---

[9] Since the spent fuel contains about 0.39 mg of plutonium per gram of uranium, the separation of 33–40 kg of plutonium a year corresponds to the processing of about 85–100 t of uranium fuel a year. For a country that has had great difficulty obtaining natural uranium for its reactor, a sustained annual requirement of about 100 t of uranium seems large. These plutonium concentration values correspond to a fuel burn-up of only about 300–400 MWth-days per tonne of uranium. See *Heavy-Element Concentration in Power Reactors*, Report no. SND-120-2 (NUS Corporation: Clearwater, Fla., May, 1977). As a result, the plutonium would contain only a few per cent plutonium-240.

[10] *Nucleonics Week*, 30 Oct. 1986.

[11] We can duplicate this estimate if we assume that the reactor operates 75% of the time and produces 1 g of plutonium per MWth-days of heat output.

Table 9.1. Estimated plutonium production in the Israeli Dimona reactor, end of 1990[a]

Years	Power (MWth)	Total Pu[b,c] (kg)	Number of warheads[d]
A: 1965–90	24	140	28
B: 1965–90	40	230	46
C: 1965–75	40	..	..
1976–90	70	330	66
D: 1965–90	70	400	80
E: 1965–70	40	..	..
1970–77	70	..	..
1978–90	150	590	118

[a] These historical estimates of the amount of plutonium produced in the Dimona reactor are based on information summarized in the text. They provide an upper and lower bound on plutonium production, although our best estimate encompasses the middle three scenarios. The A scenario is a lower bound on plutonium production and uses the lowest known power rating of the reactor, 24 MWth, throughout the period under consideration. The B scenario assumes a higher power rating of 40 MWth that is increased to 70 MWth during the mid-1970s, a time when the number of people working at Dimona is reported to have increased dramatically. The D scenario assumes a continuous power rating of 70 MWth. The E scenario represents an upper bound on plutonium production and arbitrarily assumes a step-wise increase in power from 40 MWth to 150 MWth. The latter increase is considered the theoretical maximum power achievable in this reactor.

[b] It is assumed that the reactor operates at full power an average of 60% of the time, and that it produces about 1 g of plutonium per MWth-day.

[c] Israel may have separated tritium from lithium-6 targets irradiated in the Dimona reactor. Tritium production would have reduced estimated plutonium production somewhat.

[d] It is assumed that each warhead requires 5 kg of weapon-grade plutonium (see text).

In all of these estimates, we assume that the reactor has operated reliably at an average 60 per cent capacity factor since 1965. In reality, the reactor might have experienced operating difficulties that would have lowered the capacity factor, or it might have been shut down for extended periods for maintenance or safety upgrades.

Table 9.1 summarizes our results. Our best estimate is that up to the end of 1990, Israel produced about 230–400 kg of weapon-grade plutonium. Our maximum estimate is 590 kg and our minimum estimate is 140 kg.

Production during 1991 would increase our best estimate to 240–415 kg. For future years, we assume that the power of the reactor remains between 40 and 70 MW. Up to the end of 1995, Israel could produce a cumulative total of 275–475 kg of weapon-grade plutonium; and up to the end of 2000, it could produce a total of 320–550 kg of plutonium (see table 9.2).

Estimates of future production are uncertain because the reactor is nearly 30 years-old and might be closed down soon for safety reasons. In addition, the USA is reported to have pressed Israel to cease plutonium production.[12]

[12] See, for example, Hibbs, M., 'US wants Israel to cease Dimona plutonium production,' *Nucleonics Week*, 26 Sep. 1991.

Table 9.2. Estimated inventories of Israeli weapon-grade plutonium, ends of 1990, 1991 and 1995

	End of 1990	End of 1991	End of 1995[b]
Plutonium (kg)[a]	230–400	240–415	275–475
Number of warheads	46–80	48–83	55–95

[a] These estimates have not been reduced to account for tritium production, which in any case is assumed to be small.

[b] For future plutonium production projections, it is assumed that the Dimona reactor will maintain a power of 40–70 MWth, and produce about 8.8–15 kg of plutonium per year. It is also assumed that each warhead requires an average of 5 kg of weapon-grade plutonium.

Uranium enrichment

Little is known about Israel's uranium enrichment programme. Israel is known to have pursued laser enrichment, but the status or exact purpose of this effort is unclear.[13] Some public accounts report Vanunu as saying that Israel was enriching uranium using gas centrifuges, although again there are no details about the programme.[14]

Israel's nuclear arsenal

Public estimates of the amount of plutonium used in Israeli nuclear weapons are scarce, although Vanunu stated that plutonium metal buttons were shaped into roughly spherical balls, each with a mass of about 4.5 kg. If these spheres were typical weapon components, then an estimate of 5 kg per warhead appears reasonable. (For comparison, the Trinity explosion in 1945 used 6 kg of weapon-grade plutonium, and modern US weapons average about 3–4 kg of plutonium per warhead.) Based on the plutonium production estimates in table 9.1, Israel could have constructed between 52 and 94 warheads up to the end of 1990, and could have produced another 2–4 weapons in 1991.

II. India

India has the largest nuclear programme among the developing nations. Within that programme is a sizeable nuclear weapon effort based on producing separated plutonium.

Although India continues to deny that it has deployed nuclear weapons, there is strong evidence that it could quickly deploy an arsenal of fission bombs. In 1974, India detonated a 12-kt 'peaceful' nuclear explosive. Since then, it has continued its nuclear weapon research and development programme, apparently

[13] Department of Political and Security Council Affairs, UN Centre for Disarmament, *Study on Israeli Nuclear Armament* (United Nations: New York, 1982); Report of the UN Secretary-General, 'Israeli nuclear armament', A/42/581, 16 Oct. 1987; and *Sunday Times* (note 8).

[14] See Spector and Smith (note 6), p. 161.

intensifying it in the mid-1980s in response to Pakistan's progress towards nuclear weapons.

Indian officials have often hinted at how quickly they could have nuclear weapons. In 1990, P. K. Iyengar, head of the Indian Atomic Energy Agency, said, 'In how much time we make it, will depend on how much time we get.'[15] In an interview with one of the authors in 1990, Dr Iyengar said that India could make nuclear explosives in a matter of weeks.

At the core of India's nuclear weapon programme are two small heavy water-moderated reactors and a plutonium-separation plant. The Indian civil and military programmes, however, are not clearly separated. Indian Government policy has been to minimize international safeguards on its nuclear facilities, in effect freeing them for military use if desired. As a result, most of its CANDU power reactors and its oxide spent-fuel reprocessing plant are unsafeguarded and could be used to make plutonium for weapons.

India has not constructed any large-scale uranium enrichment facilities, although it does maintain an enrichment research and development programme at the Bhabha Atomic Research Centre (BARC), near Bombay. India has also built a somewhat larger enrichment facility in the south of the country.

Plutonium production

Cirus and Dhruva

India's two largest research reactors are unsafeguarded sources of weapon-grade plutonium. They are located at BARC, and use natural-uranium fuel and a heavy-water moderator.

The 40-MWth Cirus reactor began operation in 1960. Each year, Cirus can produce about 8.8–10 kg of weapon-grade plutonium in about 8.8–10 tonnes of fuel.[16] Although the Cirus reactor was supplied by Canada on condition that it be used for peaceful purposes only, India maintained after the 1974 test that this did not preclude the use of Cirus plutonium for 'peaceful' nuclear explosives.

The 100-MWth Dhruva reactor, which was developed indigenously, went critical on 8 August 1985 at BARC. Since the Dhruva reactor is similar to the Cirus reactor, we estimate that it discharges about 22–26 kg of weapon-grade plutonium annually, assuming a capacity factor of 60–70 per cent. Iyengar, however, claimed that the annual output of Dhruva is 30 kg a year.[17]

Because of severe vibrations in the reactor core, Dhruva was shut down soon after starting.[18] In December 1986, it began operating at 25 MWth and was

---

[15] Collected scientific papers of Dr. P. K. Iyengar: Volume 5, Selected Papers and Speeches on Nuclear Power and Science in India (Bhabha Atomic Research Centre: Bombay, June 1991), p. 247.

[16] The plutonium production rate assumes the production of 1 g of weapon-grade plutonium per MWth-day of heat output and an average annual capacity factor of 60–70%.

[17] Hibbs, H., 'Indian reprocessing program grows, increasing stock of unsafeguarded Pu', Nuclear Fuel, 15 Oct. 1990.

[18] 'India's supply of unsafeguarded Pu grows as reprocessing of MAPS fuel begins', Nuclear Fuel, 11 Aug. 1986.

reported to be still operating at this level in the spring of 1987.[19] The vibrational problems were subsequently solved, and its power was increased to 80 MWth, and reached 100 MWth in mid-January 1988.[20]

CANDU reactors

India also has unsafeguarded CANDU power reactors at the Madras Atomic Power Station (MAPS) and the Narora Atomic Power Station (NAPS) that might be used to produce plutonium for weapons. But no direct evidence exists that India has done so, or plans to do so.

During normal operation, CANDU reactors produce non weapon-grade plutonium. Up to the end of 1991, MAPS-1, MAPS-2 and NAPS-1 produced over 500 kg of fuel—or reactor-grade plutonium—none of which is subject to international safeguards or restricted to peaceful purposes (see chapters 5 and 6). India would undoubtedly know how to use low-quality plutonium in nuclear weapons, but an Indian official at BARC informed a visiting US expert that India is unlikely to use reactor-grade plutonium in its weapons. India has confidence in weapon designs based on its 1974 test, which used weapon-grade plutonium. It is unlikely to want to build untested designs, if it does not have to do so.

During start-up, however, the power reactors typically produce small amounts of high-quality plutonium. During the first several months of operation, each of the MAPS and NAPS reactors could have discharged low burn-up fuel containing roughly 5 kg of high quality plutonium. This plutonium might have been assigned to the weapon programme.

India could in principle dedicate one or more of its power reactors to weapon-grade plutonium production, although with a sizeable penalty in fuel costs. Each 235-MWe reactor could in these circumstances produce up to 150 kg of weapon-grade plutonium a year, operating at 60 per cent capacity.[21]

Plutonium separation capability

India has had the capability to separate plutonium since 1964, when it commissioned the unsafeguarded Trombay reprocessing facility at BARC. The Trombay facility was shut down in 1974 for decontamination and expansion, and restarted in 1983 or 1984. Trombay is currently sized to handle fuel from both the Cirus and Dhruva reactors. According to India's Department of Atomic

[19] Chellaney, B., 'Indian scientists exploring U enrichment, advanced technologies', *Nucleonics Week*, 5 Mar. 1987.

[20] Tefft, S., 'Nuclear emergency planning delays India's first 500 MWe reactor', *Nucleonics Week*, 17 Dec. 1987; and Indian Embassy, Washington, DC, Apr. 1988.

[21] The burn-up of spent natural uranium-oxide fuel from a CANDU reactor containing 94% Pu-239 is about 1200 MWth-days/t of uranium and the spent fuel contains 1 kg of plutonium per tonne (NUS [note 9]). Therefore, about 0.83 g of weapon-grade plutonium are produced per MWth-day. Since the thermal efficiency of a CANDU reactor is about 0.28, a 1000-MWe CANDU reactor would theoretically discharge about 650 kg of weapon-grade plutonium each year at a capacity factor of 60%. It would require about 650 t of uranium per year.

Energy's annual report for 1987–88, Trombay has processed Dhruva's spent fuel.

India can also separate plutonium in the PREFRE facility, located near Bombay, which began operation in 1979. Although designed to separate plutonium from CANDU power reactor fuel, it first processed Cirus's spent fuel.[22]

The nominal annual capacity of this facility is usually listed as 100–150 tonnes of CANDU spent fuel a year. If this facility reprocesses low burn-up spent fuel, which contains roughly 1 kg of plutonium per tonne of spent fuel, it could separate at most 100–150 kg of weapon-grade plutonium a year.

Separated weapon-grade plutonium inventory

The Indian Government treats information about separated plutonium as classified, whether it is assigned to weapon or civil use. Our estimate of India's stockpile of separated weapon-grade plutonium is therefore difficult to confirm. As mentioned earlier, India also has a stockpile of fuel- or reactor-grade plutonium, which we estimate at between 100 and 200 kg at the end of 1990 (see chapter 6).

The bulk of India's weapon-grade plutonium is assumed to have been produced in the Cirus and Dhruva reactors. We can estimate this amount from the operating history of these reactors. Assuming that the Cirus reactor produced little plutonium during its first three years of operation, but operated continuously from 1964 onward, it would have produced about 245 kg of weapon-grade plutonium up to the end of 1991, assuming a capacity factor of 60 per cent. After initial start-up problems, the Dhruva reactor began operating at full power in 1988. We estimate that until 1988 the reactor produced about 10 kg of plutonium, and after that about 22 kg a year. By the end of 1991, it would have produced in total about 100 kg of plutonium.

In late 1985 or early 1986, the PREFRE facility began reprocessing spent fuel from the MAPS reactors.[23] Public information about the grade of the plutonium separated at PREFRE is unavailable, although we assume that most of the separated plutonium is fuel- or reactor-grade. The only exception is the high-quality plutonium in the fuel initially discharged from the CANDU reactors—about 15 kg up to the end of 1991 from MAPS-1, MAPS-2 and NAPS-1. Adding this amount to the inventory, India is estimated to have produced about 360 kg of weapon-grade plutonium by the end of 1991, virtually all of which is already separated.

From this estimate of weapon-grade plutonium production, we must subtract about 10 kg that were consumed in the preparation and conduct of India's 1974

---

[22] Reference to these campaigns and the starting date of the facility are in Government of India, Department of Atomic Energy, *Annual Report 1980–1981*, pp. 4 and 31; *Annual Report 1981–82*, p. 26, and *Annual Report 1983–84*, pp. 6 and 31.

[23] See *Nuclear Fuel* (note 18).

Table 9.3. Estimated inventories of Indian weapon-grade plutonium, ends of 1990, 1991 and 1995

Figures are in kilograms.

	End of 1990	End of 1991	End of 1995
Production			
Cirus reactor	235	245	280
Dhruva reactor	75	100	190
CANDU (first discharges)	15	15	30
Total production	325	360	500
Consumption			
1974 test	– 10	– 10	– 10
Processing losses (3%)	– 10	– 10	– 15
Fast reactor	– 50	– 50	– 50
Total consumption	– 70	– 70	– 75
Total	**255**	**290**	**425**

nuclear test, about 10 kg lost during separation and processing, and another 50 kg that has gone into the initial core of the unsafeguarded fast breeder test reactor. (The initial core of this reactor was fabricated before PREFRE began separating lower-quality, unsafeguarded plutonium.) About 50 kg of plutonium was also used in the Purnima research reactor, but has probably been recovered.

At the end of 1991, the weapon-grade plutonium inventory available for use in weapons was therefore about 290 kg. Assuming that India would use about 5 kg of weapon-grade plutonium in each weapon, it has enough for almost 60 nuclear weapons. We assign an uncertainty of about 25 per cent to this estimate. These results are summarized in table 9.3.

India's inventory of weapon-grade plutonium will grow mainly from continued operation of the Cirus and Dhruva reactors. By the end of 1995, India might have in total about 425 kg of weapon-grade plutonium. This includes a small amount from CANDU reactors scheduled to begin operating in the next few years.

Although unsafeguarded, Cirus plutonium is restricted to peaceful uses through agreement with Canada, which would forbid its use in nuclear weapons. This would lower the amount available for weapons up to the end of 1991 to about 115 kg, or enough for over 20 weapons. However, it is unknown whether India could substitute or has substituted an equivalent quantity of fuel- or reactor-grade plutonium for its CIRUS plutonium, thus freeing it for use in weapons. CIRUS plutonium is not safeguarded, and thus its location or use is not subject to verification.

Uranium enrichment programme

In November 1986, India stated that it could enrich uranium to whatever level it required.[24] Since at least the early 1970s, India is believed to have conducted gas-centrifuge research.[25]

The experimental gas centrifuge programme is based at BARC, where there is a pilot plant.[26] In early 1992, the Pakistani *News of Rawalpindi* said that this plant had 100 centrifuges.[27]

No official information is available about the status of its centrifuge technology, or the level and amount of enriched uranium it has produced. The experimental facility was reported in 1986 to have achieved enrichments of less than 2 per cent.[28] But Indian officials have expressed confidence that they could produce higher enrichments in whatever quantity required, although they have refused to be more specific.

India is capable of producing maraging steel, a super-strong steel ideal for centrifuges, which might imply that it can build centrifuges comparable in quality to those of Pakistan.[29] Assuming that India has managed to make a centrifuge that can produce 3 SWU per year, it could produce up to 2 kg of weapon-grade uranium per year in its experimental plant, asuming a tails assay of 0.5 per cent.

India was reported in 1986 to have decided to build a larger facility, called the Rare Materials Plant (RMP) at Ratanhalli near Mysore.[30] In an interview in March 1992 with one of the authors, Dr Iyengar confirmed the existence of this plant.[31] If the new facility had 300 centrifuges, each with a capacity of 3 SWU per year, it could produce up to 6 kg of weapon-grade uranium each year.

III. Pakistan

Pakistan has both plutonium separation and uranium enrichment programmes, both of which appear strongly oriented towards obtaining material for nuclear weapons. Its reprocessing programme has stalled because of a lack of unsafeguarded spent fuel. Its only operating power reactor is under IAEA safeguards. However, its unsafeguarded enrichment plant at Kahuta, near Islamabad, has produced weapon-grade uranium, and is the core of its nuclear weapon programme.

[24] Chellaney (note 19).

[25] 'German nozzle-enrichment knowhow could be passed to India', *Nucleonics Week*, 14 Oct. 1971, p. 8.

[26] Fera, F. and Srinivasan, K., 'Keeping the nuclear option open: what it really means', *Economic and Policy Weekly*, vol. 21, no. 49 (6 Dec. 1986), pp. 2119–20.

[27] This information about the pilot plant was included on a list of facilities that was exchanged by Pakistan and India as part of a bilateral agreement not to attack each other's nuclear facilities. Intended as secret, the document was obtained by the *News of Rawalpindi*.

[28] See Fera and Srinivasan (note 26).

[29] 'Public firm produces maraging steel', Delhi Domestic Service in English, 27 Sep. 1987.

[30] See Fera and Srinivasan (note 26); and 'The mini-superpower', *Foreign Report*, 14 Jan. 1988.

[31] Interviews in Bombay with Dr P. K. Iyengar and other Atomic Energy Department officials.

US Senator Larry Pressler, while in Islamabad in early 1992, said that 'some covert steps by Pakistan' in 1990 led the Central Intelligence Agency, the State Department and the White House to believe that Pakistan had built an atomic bomb.[32] Bush Administration officials were quoted in the *Washington Post* soon afterwards as saying that Pakistan had all the essential components for at least two nuclear weapons.

In early February, soon after Pressler's visit, Pakistani Foreign Secretary Shahryar Khan told the *Washington Post* that his nation now had the components to assemble at least one nuclear explosive device.[33] Khan said he was speaking candidly to 'avoid credibility gaps' that he suggested had been created by previous Pakistani governments. He said that his government stopped producing highly enriched uranium in 1991.

Enrichment programme

During the mid-1970s, Pakistan created a well-funded, secret procurement network in the West aimed at obtaining uranium-centrifuge enrichment technology and components, and the equipment to make the centrifuges themselves. This programme has been largely successful, and Pakistan now has an indigenous enrichment effort based at Kahuta. Although questions remain about whether Pakistan can produce all the raw materials necessary to build centrifuges, it is believed to have established a sizeable reserve of the materials necessary to maintain its centrifuge programme.

Although most experts date Pakistan's small nuclear arsenal to 1990, Pakistan is believed to have produced weapon-grade uranium in sufficient quantity for nuclear weapons several years earlier.

Pakistan's efforts to produce weapon-grade uranium

After almost a decade of concerted effort, Pakistan announced in 1984 that it was capable of producing low-enriched uranium at its pilot centrifuge plant at Kahuta.[34] By mid-1986, US intelligence had concluded that Pakistan had produced weapon-grade uranium at this facility.[35]

According to a declassified 1986 memorandum for Henry Kissinger, who received the document in his capacity as a member of the President's Foreign Intelligence Advisory Board, Kahuta had a nominal capability sufficient to

---

[32] MacLachlan, A., Shahid-ur-Rehman Khan and Siddiqi, A. R., 'Pakistani says France will pay $118 million for supply breach', *Nucleonics Week*, 16 Jan. 1992.

[33] Smith, J., 'Pakistan can build one nuclear device', *Washington Post*, 2 Feb. 1992.

[34] 'Scientist affirms Pakistan capable of uranium enrichment, weapons production', *Nawa-I-Waqt*, 10 Feb. 1984, magazine supplement, pp. F1-F8. Translated in FBIS/NDP, 5 Mar. 1984, pp. 32–45.

[35] Woodward, B., 'Pakistan reported near atom arms production', *Washington Post*, 4 Nov. 1986, p. A1; and Smith, H., 'A bomb ticks in Pakistan', *New York Times Magazine*, 6 Mar. 1988. In early 1986, Pakistan was reported to have enriched uranium to over 30%. This was the conclusion of Western intelligence agencies which analysed the uranium in dust from the immediate vicinity of the Kahuta facility. See Koch, E. and Henderson, S., 'Pakistan getting atomic bomb through back door', *Der Stern*, no. 19 (30 Apr. 1986); and 'Inside Kahuta', *Foreign Report*, 1 May 1986.

produce 'enough weapons-grade material to build several nuclear devices per year'.[36] The memorandum, however, said that Pakistan was not believed to have assembled any nuclear explosive devices.

The *New York Times Magazine* reported in March 1988 that US officials had by then concluded that Pakistan had enough weapon-grade uranium for four to six nuclear weapons.[37] This amount is roughly equivalent to 100 kg of weapon-grade uranium.

Pakistan was reported to be building another enrichment facility at the town of Golra, 6 miles west of Islamabad.[38] A 1987 report said that a centrifuge hall had reportedly been completed, although 'Western diplomats say the several thousand centrifuges needed have not yet been installed'.[39] Pakistan has reportedly proceeded with the construction of this plant, although it is not known how far it has got.

Prime Minister Benazir Bhutto was reported to have ordered a halt to any weapon-grade uranium production before her visit to Washington in June 1989, a step that the USA was able to verify prior to her visit.[40] In reaction to increased tensions with India over Kashmir, however, Pakistan decided by the spring of 1990 to resume producing weapon-grade uranium.

The Kahuta centrifuges

A. Q. Khan, the father of the Pakistani enrichment programme, is widely believed to have stolen the designs and technical data for several centrifuges from Urenco, the Anglo-Netherlands-German enrichment consortium. Khan also gained access to details about Urenco suppliers that Pakistan used in creating its procurement network. Khan had access to all this information when he worked from 1972 to 1975 at a Netherlands engineering firm whose parent company played an important role at the Urenco pilot enrichment facilities at Almelo, in the Netherlands.

Pakistan is believed to have obtained sensitive design information for early-generation German centrifuges, known as G-1 and G-2, and Netherlands prototype machines, referred to as SNOR and CNOR. According to the Netherlands Government investigative report on the Khan case, Pakistan is also believed to have acquired several SNOR and CNOR machines. These machines have a separative capacity of between 2 and 5 SWU per year.

Despite this head start, Pakistan encountered many difficulties in building reliable gas centrifuges. Its efforts to duplicate and operate the Netherlands machines evidently ended in failure after it built several thousand of them during the early 1980s. It was more successful with its machine based on the G-2

[36] The National Security Archives, *US Nuclear Non-proliferation Policy: 1945-91*, Document no. 02328, 1992.

[37] See Smith (note 35).

[38] Henderson, S., 'Pakistan builds second plant to enrich uranium', *Financial Times*, 11 Dec. 1987.

[39] See Henderson (note 38).

[40] Hussain, M., 'Nuclear issue: ball now in Pakistan's court', *The Nation*, Lahore, 29 Nov. 1990 (in English).

design, however, and by the mid-1980s Pakistan was able to mass-produce them. Pakistan's enrichment programme is believed to rely on this type of machine, although it still might be operating a certain number of older models.

The G-2 has a capacity of about 5–6 SWU a year. It contains two maraging steel rotor tubes connected by a maraging steel bellows, which acts to reduce damaging vibrations in super-critical centrifuges. Mastering the construction of bellows is likely to have presented special difficulties for Pakistan, and might be a major reason for its long delay in developing a successful machine.

In an interview in the spring of 1991, a US official with access to intelligence information on Pakistan's programme said that Kahuta had close to 3000 operating machines at any given time. At an average of 3–5 SWU per machine, Kahuta has a capability of about 9000 to 15 000 SWU per year. This is enough to produce about 45–75 kg of weapon-grade uranium a year, assuming that about 0.3 per cent of the uranium-235 remains in the waste, or tails.[41] This number would be about 30 per cent higher if a tails assay of 0.5 per cent was used.[42]

Pakistan has the manufacturing capability and know-how to increase its number of machines. But the US official said that Pakistan was concentrating on developing more advanced machines and replacing older centrifuges rather than increasing the number in operation.

The number of operating machines during previous years is highly uncertain. Part of the reason for confusion is that Pakistan has installed considerably more machines than it has successfully operated. In 1986 it was reported that Kahuta had 14 000 centrifuges.[43] US Government officials confirmed that Pakistan might have installed this number of centrifuges, but they said that a more accurate count of the number operating in 1986 or 1987 was closer to 1000. One added that Pakistan's centrifuge 'junk pile was sizeable'.

The lower estimate of the number operating in 1986 is also consistent with reports published in Pakistan by Islamabad's English daily newspaper, *Muslim*. It reported that Kahuta was 'rumored to have 1,000 centrifuges, against a planned capacity of 2000 to 3000 centrifuges'.[44]

[41] Pakistan has a uranium mining capability sufficient to supply the Kahuta enrichment plant, and it can produce uranium hexafluoride.

[42] Going to a higher tails assay requires significantly more natural uranium feed. For example, production of 25 kg of weapon-grade uranium requires 4600 kg of natural uranium at 0.2% tails assay, and 11 000 kg at 0.5% tails assay.

[43] See Koch and Henderson, 'Inside Kahuta' (note 35). The centrifuges were reportedly housed in two halls at Kahuta. One hall held centrifuges based on a Netherlands design, and made from special steel and aluminium. Each centrifuge stands about 2.5 m tall and is about 11 cm in diameter. The other hall was reported to hold centrifuges based on the German design, which are made of maraging steel, are only half as tall as the Dutch model and are 15 cm wide.

[44] Ameen, A. F., 'The mythical bomb', *Muslim*, 5 Aug. 1986. Quoted in Khan, S., 'Fear of US aid cutoff said to have deterred Pakistan's bomb program', *Nuclear Fuel*, 11 Aug. 1986.

Table 9.4. Estimated production of weapon-grade uranium at the Pakistani Kahuta centrifuge enrichment plant, 1986–91

| Year | Capacity (SWU/yr) | Weapon-grade uranium | |
		Annual production[a] (kg)	Cumulative production (kg)
1986	3 000–5 000	15–25	15–25
1987	4 500–7 500	23–38	38–63
1988	6 000–10 000	30–50	68–113
1989	7 500–12 500	15–25[b]	83–138
1990	9 000–15 000	24–40[b]	107–178
1991	9 000–15 000	23–38[c]	130–216

[a] Tails assay of 0.3% and natural uranium feed.

[b] It is assumed that no weapon-grade uranium was produced from the beginning of May 1989 until June 1990 as a result of a moratorium instituted by then Prime Minister Benazir Bhutto (see text).

[c] Assumes weapon-grade uranium production in the first half of 1991 only.

Inventory of weapon-grade uranium

Table 9.4 gives an estimate of the annual production of weapon-grade uranium at Kahuta, based on the above information. Up to the end of 1991, Pakistan is estimated to have produced about 130–220 kg of weapon-grade uranium, using natural uranium feed and a tails assay of 0.3 per cent.

In this estimate, we assume that weapon-grade uranium production began in earnest in 1986, and continued uninterrupted until 1989. After about a one-year pause, production resumed, stopping again in 1991. We also assume that Pakistan linearly increased Kahuta's enrichment capability during this period from an initial capacity of 1000 machines in 1986 to 3000 machines in 1990, each with an average capacity of 3–5 SWU per year. This estimate of weapon-grade uranium production is highly uncertain, and we therefore assign an uncertainty of plus or minus 25 per cent.

There have been reports that Pakistan obtained enough weapon-grade uranium from China to make two nuclear bombs, although the authenticity of this information cannot be determined.[45] These reports cannot be dismissed, however, because a companion report that China supplied Pakistan with a nuclear weapon design has been confirmed to one of the authors by US officials on several occasions. But without confirmation of the weapon-grade uranium, we do not add this amount to our estimate of Pakistan's inventory.

Little specific information is available about the amount of weapon-grade uranium required in the bomb design supplied by China. A US official said that

[45] 'Pakistan's atomic bomb', *Foreign Report*, 12 Jan. 1989. Pakistan is thought to have acquired about 1 kg of weapon-grade uranium as part of a deal with China for the supply of a 27 kW reactor. This reactor, at the Pinstech nuclear centre, went critical in 1989. It is under IAEA safeguards, precluding its military use.

the design provided by China has a solid core of weapon-grade uranium. This type of design usually does not require more than 20 kg of weapon-grade uranium. Under this assumption, at the end of 1991, Pakistan had enough material for roughly 6–10 nuclear weapons.

This estimate is consistent with a late-1990 report in the *Press Trust of India* which quotes a senior Indian official as saying that Pakistan had nine nuclear bombs.[46] In September 1991, US officials asserted that Pakistan had the parts and capability to assemble up to six nuclear bombs.[47]

If Pakistan resumes its production of weapon-grade uranium, it could accumulate enough over the next few years for several additional nuclear weapons. If Pakistan operates 3000 centrifuges arranged to produce weapon-grade uranium from natural uranium, each with a capacity of 5 SWU per year, it could produce 75 kg of weapon-grade uranium each year. At this rate, its weapon-grade uranium stockpile in 1995 could reach 430–520 kg, or enough for roughly 20–25 atomic bombs, assuming production resumed in early 1992.

Plutonium separation

Despite a lack of spent fuel free of safeguards and foreign control, Pakistan has established a small capability to separate plutonium. A US official interviewed in 1991 said that Pakistan had finished a small reprocessing facility, called 'New Labs', at the Pinstech facility, near Rawalpindi. The capacity of the facility is believed to be small, although it could be expanded.

The Chashma plutonium separation plant, a medium-size reprocessing plant able to process 50–100 tonnes a year of spent CANDU fuel, was to be supplied by France in the mid-1970s. France, however, decided not to fulfill the contract in 1977, under pressure from the United States.

According to an article in the *Muslim*, New Labs is based on blueprints delivered by France before it broke its contract to supply the Chashma reprocessing plant.[48] Key plant equipment was evidently bought clandestinely.

[46] 'Indian politician claims Pakistan has 9 nukes', *Washington Times*, 27 Nov. 1990.
[47] Levine, S., 'Bhutto says Pakistan can build nuclear weapons,' *The Guardian*, 2 Sep. 1991.
[48] See Khan (note 44).

10. Countries of concern: Iraq, North Korea, Iran and Algeria

Several countries are widely suspected of wanting to acquire nuclear weapons and of actively developing the capability to make plutonium or highly enriched uranium outside international safeguards. Iraq, North Korea and Iran are currently highest on this list of suspect nations, even though they are parties to the Treaty on the Non-Proliferation of Nuclear Weapons (NPT). Algeria has conducted secret nuclear activities that have heightened concerns about its ultimate ambitions, although these concerns have been diminished by its signing a safeguards agreement for a new research reactor.

Other countries, notably Libya and Cuba, have at various times been suspected of wishing to develop the capability to make nuclear explosive materials. This study does not consider these countries because they do not appear to have made significant progress to date.

South Korea took steps to acquire nuclear weapons in the 1970s, but as a result of international pressure its activities in that direction were halted. Currently, it is not suspected of having a nuclear weapon programme, and is thus not considered here. However, its activities will be closely watched as long as North Korea's programme continues.

A special case is Iraq. Under UN Resolution 687, adopted on 3 April 1991, Iraq's nuclear, chemical and biological weapons are being systematically destroyed. However, there is concern that Iraq may try to resurrect its nuclear programme if given the chance, especially if the Saddam Hussein Government remains in power.

I. Iraq

The full extent of Iraq's nuclear programme was not revealed until after the Persian Gulf War. Under a 'go-anywhere, see-anything' mandate from the United Nations, International Atomic Energy Agency (IAEA) inspectors have systematically uncovered Iraq's nuclear weapon programme, including an enrichment programme that might have been only two to four years away from producing significant amounts of HEU.[1] The IAEA's major findings are

---

[1] Albright, D. and Hibbs, M., 'Iraq's nuclear hide and seek', *Bulletin of the Atomic Scientists*, vol. 47, no. 7 (Sep. 1991), pp. 14–23; Albright, D. and Hibbs, M., 'Iraq's bomb: blueprints and artefacts', *Bulletin of the Atomic Scientists*, vol. 48, no. 1 (Jan./Feb. 1992), pp. 30–40; Albright, D. and Hibbs, M., 'It's all over at Al Atheer', *Bulletin of the Atomic Scientists*, vol. 48, no. 5 (June 1992), pp. 8–10.

presented in a series of reports available from the United Nations. This section is largely based on these reports.

Iraq's nuclear know-how is so extensive that its activities and personnel will require long-term monitoring by the UN and the IAEA. Without international constraints, Iraq could resume its nuclear weapon programme.

The Iraqi enrichment programme

The Iraqi enrichment programme appears to have been accelerated following the Israeli bombing of the Osirak reactor. Israel alleged that this reactor would become a source of plutonium for nuclear weapons. Iraq's subsequent nuclear activities have made this charge more plausible.

Iraq worked on at least four different enrichment technologies: gaseous diffusion, chemical, electromagnetic and gas-centrifuge enrichment. The last two were being developed into production-scale operations when the Persian Gulf War began and the programme was halted.

The calutron programme

Iraq's most developed enrichment technology was based on calutrons, in which an intense source of uranium ions is isotopically separated by a magnetic field (see chapter 2). Calutrons were probably the most accessible enrichment technology to Iraq, despite the technology's high cost and inefficiency.

Iraq carried out a large research and development effort on all aspects of calutrons at the Al Tuwaitha nuclear research centre near Baghdad. During the development phase, several separators of different designs were built. In the early 1980s, Iraq built a small machine with a beam-current of 1 milliampere. Over the next several years, it scaled up its machines. Before the war, it was building production models containing four ion sources, with total design currents of 400–600 milliamps. In parallel with separator development, Iraq developed chemical processing at Al Tuwaitha to recover the uranium from the calutrons.

During the research and development phase at Al Tuwaitha, which lasted until 1990, the highest enrichments achieved were 17 per cent for gram quantities and 45 per cent for milligram quantities.[2] Results from environmental samples taken at Al Tuwaitha are consistent with these declared enrichments.

The next phase of the programme aimed at reaching industrial-scale production and started with the February 1990 opening of a facility at Tarmiya, 40 km north of Baghdad. During the following nine months, Iraqi scientists installed eight large 'alpha' calutrons, designed to produce low-enriched uranium (LEU) from natural uranium, and operated them at declared average capacities of 15 per cent. They produced about 500 grams of uranium, averaging 4 per cent enrichment, with a maximum of about 10 per cent. The chemical processing

[2] *Report on the Fourth IAEA On-Site Inspection in Iraq under Security Council Resolution 687 (1991), 27 July –10 August 1991*, United Nations Security Council document S/22986, 28 Aug. 1991, p. 6.

was done at Al Tuwaitha, since the new chemical facilities at Tarmiya were not finished at the time.

The Tarmiya facility was planned as a production facility, but it was functioning as an advanced research and development facility for calutrons when it was bombed by the Allies in January 1991. Iraq was then beginning to install 17 other alpha machines, although not all the components for them had been manufactured. A total of 70 alpha machines was planned.

The LEU produced in the alpha machines would have been enriched further, to weapon grade, in 20 smaller 'beta' machines, six of which were under construction when the Tarmiya facility was bombed. None of these had been installed at the time of the bombing, and Iraq had not yet demonstrated an ability to enrich uranium up to weapon grade.

Iraq was building a twin facility at Al Sharqat, 200 km north-west of Baghdad, which was about 80–90 per cent complete when it was bombed. It is unclear whether Iraq intended to finish this plant; according to declarations it did not. In any event, no calutrons had been installed there.

Experts differ on the output that the Tarmiya facility could have achieved. The transition from prototype machine operation to industrial production can be difficult, and extrapolations are at best fraught with uncertainty. At most, if the machines were to operate steadily at optimal design currents and 100 per cent capacity, output could reach as high as 20–27 kg of weapon-grade uranium per year. It is doubtful if Iraq could have sustained such optimal conditions; more likely production would have been less than 15 kg of weapon-grade uranium per year.

The lowest estimate is contained in a 1987 report by an Iraqi calutron expert. The report was seized by inspectors in autumn 1991 and translated several months later. This expert concluded that once all 90 calutrons were installed, the plant would produce about 7 kg of weapon-grade uranium per year, using natural uranium feed.

The biggest unknown is how long it would have taken Iraq to reach industrial-scale operation. The IAEA's fourth inspection report said: 'It is possible—but by no means certain—that full production operation at Tarmiya might not have been achieved for another 18–36 months'.[3] The Iraqis cited major problems in producing enough uranium tetrachloride feedstock, and in supplying graphite for the uranium collectors. The fourth report also states that 'there may have been human resource problems associated with these large facilities'.[4]

The 1987 Iraqi report referred to above said that installation of all 90 machines would have taken about 40 months. Allowing for delays in installing them before the Persian Gulf War, installation, following the plan laid out in the report, would have been complete early in 1994.

[3] United Nations (note 2), p. 8.
[4] United Nations (note 2), p. 8.

The gas centrifuge programme

Iraq probably started its gas centrifuge programme at about the same time as its calutron programme, but its effort to master high-speed centrifuges progressed more slowly. Since centrifuges are more efficient, reliable and easily hidden, Iraq probably viewed them as ideal replacements for calutrons.

The ninth IAEA on-site inspection team concluded that Iraq's centrifuge enrichment programme had not progressed to a point where they could have started a sizeable production of centrifuges, although, given time, they would probably have been successful.[5] As in the case of calutrons, the centrifuge programme might have faced serious problems as it moved from laboratory testing of gas centrifuge equipment to industrial production of significant quantities of enriched uranium.

Information to date shows that Iraq obtained significant foreign assistance. Whether it had received enough assistance prior to the Persian Gulf War to master centrifuge manufacturing or whether on-going assistance would have been necessary remains unclear. If continued assistance was essential, then its foreign procurement efforts might have been significantly hindered by more effective export controls being implemented in Germany and elsewhere.

By the time of the war, Iraq had tested only single centrifuges, and produced only minute quantities of enriched uranium. Iraq planned to develop centrifuges that use super-strong maraging steel. Machines with rotors made of this type of steel were evidently to be the backbone of the Iraqi centrifuge programme. However, at the time of the US bombing Iraq was learning to make suitable maraging steel rotors, and used a limited number of carbon-fibre rotors for its research programme. The latter type of rotor is considered more sophisticated, but also more difficult to manufacture. In 1989 and 1990, Iraq illegally obtained 22 carbon-fibre rotors from a German firm.

The capacity of its test machine had reached 1.9 SWU per year by the end of December 1990. With additional machine optimization, Iraq could have obtained a separative output of 2.7 SWU per year for its maraging steel production model.

Iraq has admitted, and the IAEA has verified, that it had obtained 100 tonnes of maraging steel, although the Iraqis have refused to reveal their foreign suppliers of the steel. This was probably 350-grade material, ideal for centrifuges. This is enough steel to produce 2500 rotors, assuming a 50 per cent reject rate during manufacturing. The IAEA received information from a Western government that Iraq may have received 400 tonnes of this grade of maraging steel. During the eleventh IAEA inspection, Iraq denied receiving more than 100 tonnes. Investigations continue.

Assuming that Iraq would have built 2500 machines, and that its machines each had a capacity of 2.7 SWU per year, the total capacity of a centrifuge plant

---

[5] International Atomic Energy Agency, 'Ninth IAEA inspection team verifies Iraqi stockpile of centrifuge parts and materials', Press Release (IAEA: Vienna, 17 Jan. 1992).

would have been 6750 SWU per year. Assuming that the tails assay was high (about 0.5%), Iraq could have produced about 40 kg of weapon-grade uranium per year, enough for about two nuclear weapons per year. Iraqi nuclear devices would have required about 18 kg of weapon-grade uranium each.

Iraqi experts said that they intended to have built and started operation of 100 machines connected together by aluminum pipes in a 'cascade' in mid-1993, and a 500-machine cascade by early 1996. Iraq might have first needed to build and operate a smaller cascade, with perhaps 10–15 machines, before constructing cascades of 100 machines or more. No evidence of a smaller cascade, however, has been found.

Other enrichment technologies

Iraq also investigated chemical exchange isotope separation and gaseous diffusion. The IAEA inspectors concluded that chemical enrichment work had not progressed very far.

Early in its enrichment programme, Iraq conducted a serious feasibility study of gaseous diffusion, including some laboratory work on barrier material. Gaseous diffusion was abandoned because it would require a huge investment, be technically demanding with little prospect of outside assistance, and be highly visible to outside intelligence agencies when completed.

Iraq's plutonium separation capabilities

Iraq separated about 5 grams of plutonium at a small laboratory at Al Tuwaitha. The plutonium came from uranium irradiated in the safeguarded IRT-5000, a small Soviet-supplied research reactor.

During the 1970s, Iraq pursued the development of a plutonium separation capability. Although the Israeli bombing of the Osirak reactor in 1981 effectively ended that quest, Iraq has since continued to learn how to separate plutonium. There is no evidence so far that Iraq was trying to enlarge the programme by secretly building a plutonium production reactor and a larger plutonium separation facility. The tenth IAEA on-site inspection looked for an underground reactor, but did not find any evidence of one, although the Agency is continuing its search.

Projected weapon-grade uranium inventory for Iraq

Iraq's nuclear weapon programme has largely been destroyed. As can be seen in table 10.1, it might have gained enough material for two atomic bombs by the end of 1995, and enough for 15 bombs by the end of the century, had the Persian Gulf War not intervened. This is based on the highly optimistic assumptions that:

Table 10.1. Maximal projection of Iraqi weapon-grade uranium inventories

Weapon-grade uranium figures are in kilograms; figures for potential bombs are cumulative.

Year	Annual production of weapon-grade uranium				Potential bombs
	Calutron	Centrifuge	Total	Cumulative[a]	
1991	..	..	..	..	..
1992	3	..	3	3	–
1993	7	2	9	12	–
1994	10	2	12	24	1
1995	10	8	18	42	2
1996	10	17	27	69	3
1997	10	40	50	120	6
1998	10	40	50	170	9
1999	10	40	50	220	12
2000	10	40	50	270	15

[a] Cumulative totals are rounded to two significant figures.

1. The calutron programme could have operated successfully without detection for at least a decade, at an average output of 10 kg of weapon-grade uranium per year.

2. Iraq could have operated the 100-machine centrifuge cascade by early 1993, increased that number to 500 in early 1995, added another 500 machines by 1996, and reached 2500 operating machines by early 1997.

II. North Korea

North Korea is believed to have pursued a nuclear weapon programme since the late 1970s. It is developing the capability to produce separated plutonium, and has been suspected of developing uranium enrichment technologies, but no indications of such a programme have been found.

At a research centre near the city of Nyongbyon, about 100 km north of Pyongyang, North Korea is operating a 5 MWe (20–30 MWth) reactor and building a plutonium separation plant. If completed, this facility might be able to process hundreds of tonnes of irradiated reactor fuel annually.

Until 1992 North Korea refused to allow IAEA inspections of this reactor, thereby violating its obligation to accept full safeguards as a party to the NPT, to which it acceded in 1985. As a result of prolonged diplomatic pressure, North Korea signed a safeguards agreement with the IAEA in January 1992. It also signed a treaty with South Korea under which neither will build nuclear weapon installations or plutonium separation plants, and which will allow bilateral inspections.

However, despite these positive developments, governments in Asia and elsewhere remain concerned that North Korea will not live up to its commitments. By early summer 1992 North Korea appeared to be backing away from its commitment not to separate plutonium. A North Korean capability to produce separated plutonium would be provocative and would create additional instability in the region. The concern is that South Korea might feel compelled to develop its own nuclear weapon capability, a step it abandoned in the mid-1970s. If this happened, Japan might decide to respond in kind.

The Nyongbyon reactors

North Korea may now be operating its small plutonium production reactor close to full power, producing enough plutonium in irradiated fuel for roughly one nuclear explosive per year. Construction of this reactor is believed to have started in 1980 and low-power operation to have begun in 1987.[6] The reactor is graphite-moderated, gas-cooled and based on 1950s technology.

The reactor experienced start-up problems, although in 1991 it was operating at about 20–30 MWth. At this power rating and assuming that it operates 60 per cent of the time, the reactor would produce about 4–6 kg of weapon-grade plutonium per year.

A larger graphite-moderated reactor, with a power level of 50 MWe (200–300 MWth), is under construction nearby. It is expected to be finished in 1995. A reactor of this type and size could produce roughly 40–60 kg of weapon-grade plutonium per year. At Taechon, North Korea is building a 200 MWe (roughly 1000 MWth) reactor of the same design. The reactor is scheduled to be finished in 1996, and could produce up to 200 kg of plutonium per year, if the fuel has a low burn-up.

Plutonium separation

At Nyongbyon North Korea is building a plutonium separation plant which is 180 metres long and several stories high. This facility is estimated to be capable of processing up to several hundred tonnes of spent fuel annually when fully operational. This is enough to handle all the fuel from all three reactors described above, which together might discharge as much as 500 tonnes of spent fuel annually, containing up to 250 kg of weapon-grade plutonium.[7]

About 80 per cent of the civil engineering is completed, but only about 40 per cent of the equipment has been installed. North Korean officials told IAEA officials during a pre-inspection visit in May 1992 that the missing equipment has been ordered, but they were unclear about when or if the plant would actually be completed. As a result, no firm estimates can be made of when the plant might be finished. However, most experts conclude that the plant might

---

[6] So Yong-ha, 'Capacity for nuclear weapons development' [in Korean], *Hoguk* (Seoul), July 1989, pp. 119–22. English translation in Foreign Broadcast Information Services, FBIS-EAS-89-148, 3 Aug. 1989, pp. 23–26.

[7] It is assumed that each tonne of spent fuel contains 0.5–1.0 kg of weapon-grade plutonium.

start operating in 1995–97, at an estimated cost of several hundred million dollars.

North Korea has declared to the IAEA that in 1990 the separation plant separated gram quantities of plutonium from damaged fuel elements removed from the 5-MWe reactor. North Korea also admitted that it had separated smaller quantities of plutonium in laboratory-scale processing equipment ('hot cells') in the capital at Pyongbyon, that was supplied by the USSR in the 1960s. It is also possible that the country may be operating a small, pilot separation plant, perhaps at a site which has escaped detection. Detection of such a facility would be difficult, and intelligence agencies have so far no evidence that it exists. North Korea denies that such a plant exists.

Projected plutonium inventory for North Korea

By the end of 1991 the small production reactor is estimated to have produced about 5–10 kg of weapon-grade plutonium. At the end of 1995, the amount could increase to about 20–35 kg of weapon-grade plutonium. When the larger reactor operates, North Korea could produce up to 250 kg of weapon-grade plutonium per year.

If the separation plant at Nyongbyon were to be completed, its large size implies that it could quickly process a backlog of spent fuel. North Korea could accumulate enough plutonium for roughly 4–7 nuclear weapons by the end of 1995. If the other two reactors are completed on schedule, the country could accumulate several hundred kg of separated plutonium by the year 2000.

III. Iran

Because it is suspected of having nuclear weapon ambitions, Iran is under a virtual Western embargo of nuclear equipment and technology. Germany has refused to help Iran finish the Bushehr nuclear power reactors, and France is resisting supplying Iran with LEU from the Eurodif enrichment plant, despite the fact that Iran retains an indirect share in the plant.

In the USA, Director of Central Intelligence Robert W. Gates testified to Congress in mid-January 1992 that Iran has 'embarked on an across-the-board effort to develop its military and defense industries'. This effort, he said, 'includes programs in weapons of mass destruction'. Towards this goal, Iran 'continues to shop Western markets for nuclear and missile technology', he said.[8] Given the poor state of relations between the USA and Iran, comments by US Central Intelligence Agency (CIA) officials have to be treated with considerable caution. Nevertheless, Gates is reflecting quite widespread concern that Iran might follow Iraq in attempting to mount a clandestine nuclear weapon programme.

[8] Quoted in Hoffman, D., 'Iran's drive to rebuild seen posing new challenges to West', *Washington Post,* 2 Feb. 1992.

Unlike the case of North Korea, Western intelligence agencies have not conclusively identified secret nuclear facilities in Iran. Media reports about uranium enrichment plants and other nuclear facilities in Iran are highly speculative. If any such facilities exist, however, it is important that they should be identified. Iran is a party to the NPT and must therefore accept IAEA safeguards on all its activities involving nuclear materials. The IAEA also has the right to request 'special inspections' to increase confidence that everything is being properly safeguarded.

To help clarify the situation in Iran, IAEA officials visited six sites during February 1992 which the IAEA itself had selected. It reported that the activities at the sites were consistent with peaceful uses. Three of the sites visited had not been seen before by IAEA inspectors. Their selection had been based on intelligence and media reports. An IAEA press release about the visit, issued on 15 February, said that 'the team's conclusions are limited to facilities and sites visited by it and are of relevance only to the time of the team's visit'.[9] This statement implies that additional inspections of this type may be sought.

US officials believe Iranian researchers are now pursuing activities related to the nuclear fuel cycle, but there is no hard information indicating that Iran has made significant progress. They cite evidence that Iranian nuclear scientists who left after the Shah was overthrown are returning to Iran. Some may have expertise relevant to uranium enrichment, and others experience relevant to plutonium separation.

Iran's plutonium separation capabilities

Although Iran does not have an unsafeguarded reactor able to produce significant quantities of plutonium, it does have 'hot cells' capable of separating gram quantities of plutonium from irradiated fuel. When the USA provided Iran with a 5-MWth research reactor in the 1960s, it also provided it with 'hot cells'.

Iran's research reactors are under safeguards. There is nevertheless concern that it might follow Iraq's example and use its reactors to produce small amounts of plutonium that could be separated in the hot cells. Although Iran could not separate enough plutonium for a bomb in this way, it could gain valuable experience in plutonium separation.

Iran's uranium enrichment capabilities

Iran has obtained a single calutron from China that was part of a deal with China for the supply of a small, 27-kW reactor to be located at Isfahan, 130 km south of Tehran. The purpose of the calutron is to produce stable isotopes that would be irradiated in the reactor and converted into radioactive materials useful in research and medicine.

---

[9] International Atomic Energy Agency, Press Release (IAEA: Vienna, 15 Feb. 1992).

The machine has a current of only 1 milliampere. Although in theory capable of producing only minuscule amounts of enriched uranium, an operational 1-milliampere calutron could be useful in the development of larger machines. According to the report on the fourth IAEA on-site inspection in Iraq, in the first stage of Iraq's calutron programme a 1-milliampere machine was built that 'permitted the testing of insulator and liner concepts'.[10] The small calutron could also be reverse engineered and expanded in size.

However, Iran would first need to put together an industrial infrastructure in order to make the necessary components, or find a way to buy them abroad. In an interview with one of the authors, a US official expressed the opinion that Iran is a long way from either prospect.

According to unconfirmed media reports, Iran may also be pursuing gas centrifuge development at secret locations in northern Iran, including Maollem Kalayeh, in the Elbruz Mountains, and Karaj, north of Tehran. The precise status of these facilities appears to be unknown. When IAEA officials visited Maollem Kalayeh in February 1992, they found a motel-sized retreat and language study centre for nuclear scientists.

Iran is also working on a laser uranium enrichment project that was begun under the Shah, but it is considered unlikely that the country will make significant progress in this area.

Estimates of when Iran might be able to produce fissile material are difficult to make, since so little is known about Iran's nuclear programme. If Iran has weapon intentions, it may be able to produce separated plutonium or HEU by the end of this century.

IV. Algeria

In April 1991 the press reported that Algeria was secretly building a Chinese-supplied reactor at Ain Oussera, a remote site in the Atlas Mountains.[11] Soon after the presence of the reactor was revealed, the Algerian Government agreed to place it under IAEA safeguards, and a safeguards agreement was signed in late February 1992.

Algeria is not under any international obligation to divulge the existence of its nuclear facilities, or to put them under IAEA safeguards (unless required to do so by suppliers of its materials and equipment), since it is not a party to the NPT. Nevertheless, the Algerian Government's lack of candour about this reactor, even after its existence was exposed by the media, has raised concerns about Algeria's long-term intentions. The recent safeguards agreement has removed most of these concerns. On-site inspections, expected as soon as the spring or summer of 1992, could eliminate them.

---

[10] United Nations (note 2), p. 5.
[11] See, for example, Gertz, B., 'China helps Algeria develop nuclear weapons', *Washington Times*, 11 Apr. 1991, p. A7.

The Salam reactor

The reactor, which China agreed to provide in the early 1980s, is expected to be operational soon. It is reported to have a power of 10–15 MWth, to use LEU fuel, and to be moderated and cooled by heavy water. A 15-MWth reactor of this type is theoretically capable of producing about 2 kg of weapon-grade plutonium per year.[12]

Early press reports said that the power level might go as high as 45–60 MWth, based on analysis of the reactor's cooling towers. However, this claim has been discounted.

Plutonium separation

There have been no confirmed reports that Algeria has also acquired plutonium separation capabilities at the Ain Oussera site. According to David Kyd, Public Information Director at the IAEA in Vienna, the IAEA is not aware of any such facilities at the reactor site, but will not know for sure until Algeria provides detailed design information for the reactor and the site is inspected.[13]

---

[12] It is assumed that the reactor uses about 3% enriched uranium fuel, operates 70% of the time and produces about 0.6 g of weapon-grade plutonium per MWth-d.

[13] Personal communication, 22 Jan. 1992 and 13 May 1992.

11. Countries backing away from nuclear weapons: Argentina, Brazil, South Africa and Taiwan

I. Introduction

Just a few years ago, the conventional wisdom was that threshold countries were unlikely to back away from their programmes to produce unsafeguarded fissile materials. But several countries have recently shown that proliferation is reversible.

In 1990, Brazil took concrete steps to unveil and roll back its long-secret nuclear weapon programme, and agreed to permit IAEA inspections of all its nuclear facilities. In response, Argentina also agreed to do the same. In late 1991, Argentina and Brazil signed a specific agreement with the IAEA to allow these inspections.

In the summer of 1991, South Africa acceded to the NPT, and agreed to place all its nuclear materials in the country under IAEA safeguards. South Africa had the largest unsafeguarded enrichment programme outside the NWS and had acknowledged its capability to make nuclear weapons.

A less noticed development is the removal of a stock of weapon-grade plutonium from Taiwan, a country long thought to harbour ambitions to develop nuclear weapons. Although Taiwan was unable to develop a capability to separate the plutonium, the US Government was worried that it eventually would, and pressed Taiwan to send the spent fuel to the United States. This effort started in 1985, and was nearly completed at the end of 1991.

II. Argentina and Brazil

Argentina and Brazil signed a safeguards agreement with the IAEA on 13 December 1991 that commits them to international inspections of all their nuclear facilities. Two days earlier, the parliaments of both countries ratified a bilateral inspection agreement that forms the basis for the IAEA inspections.

These agreements are the result of several steps the two governments have taken since 1987 to signal their commitment to eliminating any hint of nuclear weapon programmes. Perhaps the most dramatic was in September 1990, when Brazilian President Fernando Collor de Mello closed a potential nuclear explosive test site in the Amazon, and cancelled a secret 15-year-old atomic bomb project, named 'Solimoes' after a river in the Amazon.

The IAEA safeguards agreements will apply to both countries' uranium enrichment facilities, a concession that was opposed by some members of their nuclear establishments and the Brazilian military.[1] Safeguarding these particular facilities is important because they are the main ones that could have produced fissile material for nuclear explosives. Brazil also has a small jet-nozzle enrichment plant supplied by Germany, but it has always been under IAEA safeguards and is not considered further in this chapter.

Both governments have stated as a matter of policy that they will not produce highly enriched uranium in their enrichment plants. How this pledge would be verified by IAEA or bilateral safeguards is not known.

Argentina's secret enrichment facility

Argentina surprised the world with its November 1983 announcement that for the previous five years it had been secretly building an unsafeguarded gaseous diffusion uranium enrichment facility in the hamlet of Pilcaniyeu in the Rio Negro province. At the time, the announcement revealed a failure of intelligence gathering.

The primary purpose of the Pilcaniyeu facility is to produce up to 20 per cent enriched uranium for use in domestic research reactors and for export, as well as slightly enriched uranium for power reactors. It might also be used to make naval reactor fuel, if Argentina ever developed a nuclear-powered vessel. When the existence of the plant was announced in 1983, Argentine nuclear officials expected the plant would soon reach its full capacity of 20 000 SWU per year. This would correspond to the production of 500 kg of 20 per cent enriched uranium.

The plant has been beset with delays because of severe budget cut-backs and technical difficulties. It has had problems with short barrier lifetimes, leaking seals, and compressor reliability. On numerous occasions during the past several years, Argentine officials have declined to state when Pilcaniyeu would reach full capacity, or the amount of enriched uranium already produced. Faced with delays at the plant, Argentina ordered 150 kg of 20 per cent enriched uranium from the Soviet Union. Earlier, it had obtained an unspecified amount of unsafeguarded 20 per cent enriched uranium from China. In the long term, Argentina plans to expand the capacity of the plant to 100 000 SWU per year. However, senior Argentine officials have said in interviews that this expanion is not a priority.

Despite the technical problems, Argentina has produced low-enriched uranium at Pilcaniyeu. However the information available is insufficient to estimate total production or to determine the amount of unsafeguarded enriched uranium imported by Argentina.

---

[1] Hibbs, M., 'Brazil's military may block safeguards with Argentina', *Nucleonics Week*, 28 Nov. 1991.

Production of highly enriched uranium

Argentine nuclear officials involved in building and operating Pilcaniyeu said in an interview with one of the authors in 1988 that the plant was designed to produce only up to 20 per cent enriched uranium, not highly enriched uranium. Nevertheless, the officials conceded that producing highly enriched uranium in such a plant is possible, although time-consuming. The fact that highly enriched uranium could be produced is the major reason why thorough inspections are needed at the plant.

Gaseous diffusion technology is not very flexible, since each enrichment stage increases the fractional amount of ^{235}U only slightly and requires a large volume of uranium feed. Enrichment stages are connected in a long series or cascade. The higher the enrichment, the more stages are needed. For example, producing 20 per cent enriched uranium from natural uranium would require about 2000 stages, while producing 90 per cent would require almost twice as many stages.

Since a gaseous diffusion plant requires a very long cascade, the construction of enough additional stages to make weapon-grade uranium in a plant configured to make 20 per cent enriched uranium would be difficult to do quickly, and would, therefore, be relatively easy to detect through IAEA inspections of the cascade hall.

A gaseous diffusion cascade contains a large amount of uranium gas, which severely complicates another potential route to highly enriched uranium, namely 'batch recycling'. Here, the enriched end-product of the cascade is collected and reintroduced into the cascade at the beginning. In a cascade designed to produce 20 per cent enriched uranium, only one recycle would be necessary to produce 80 per cent enriched uranium. However, the in-process inventory of a plant such as Pilcaniyeu could easily reach 2000 kg, at its planned capacity of 20 000 SWU per year. Producing this much 20 per cent enriched uranium feed in Pilcaniyeu would require four years of operation at full capacity and at a tails assay of 0.3 per cent.

The most practical way to produce highly enriched uranium at Pilcaniyeu is to 'stretch the cascade'. In this method, the enriched uranium product is extracted from the cascade at a greatly reduced rate, causing the enrichment level of the final product to increase dramatically, although at the cost of reduced efficiencies and production rates.[2] In this way, small amounts of uranium enriched to 90 per cent can be produced in a plant configured to produce 20 per cent enriched uranium. Usable amounts of roughly 80 per cent enriched uranium can be obtained by withdrawing product at a fraction of the

---

[2] The maximum enrichment level obtainable occurs when no product is removed, where the separative work is zero and the tails concentration reaches its limiting value, which is the feed concentration. The location of the feed stage moves down the cascade as the product withdrawal rate decreases and the product concentration increases—finally the entire cascade becomes an enricher. The limiting case is called 'total reflux'. The enrichment level of the product can reach 90% ^{235}U in a cascade configured to produce 20% enriched uranium using natural uranium feed. If 20% ^{235}U feed is used instead, the maximum enrichment level will exceed 99% ^{235}U.

designed value. Estimating the exact amount would require detailed knowledge of the cascade design, which is not available. However, assuming a reduction in output efficiency to 10 per cent, or 2000 SWU per year, the Pilcaniyeu plant could produce in the order of 10 kg a year of roughly 80 per cent enriched uranium.

In both batch recycling and stretching, the operator of a gaseous diffusion plant would have to take special precautions to prevent the possibility of a nuclear criticality accident in the enrichment stages and when using uranium withdrawal equipment. In particular, the operator would have to monitor the process very carefully to ensure that high-assay material does not solidify inside the equipment.

Argentina's plutonium separation programme

In 1978, Argentina announced that it would build a small reprocessing facility at the Ezeiza Research Complex near Buenos Aires. The plant has suffered many delays, and may never operate. At full capacity, the plant could separate 15 kg of plutonium per year from spent fuel from heavy water reactors. The plutonium would be recycled into heavy water reactors (see chapter 6).

Since the reprocessing facility could have produced unsafeguarded plutonium, Argentina was suspected in the late 1970s and early 1980s of planning to build an unsafeguarded plutonium production reactor. This suspicion was never substantiated and the Ezeiza facility was gradually accepted as part of a civilian research and development programme.

If this plant were operated successfully, Argentina might build a larger reprocessing plant at the end of this century. It envisages using the plutonium in either its heavy water reactors or breeder reactors, which it still hopes to develop in conjunction with Brazil.

Brazil's secret enrichment programme

The centre-piece of Brazil's autonomous nuclear programme is the Aramar gas centrifuge enrichment plant, operated by the Navy near Sorocaba in the state of São Paulo. Like Pilcaniyeu, it has been slow to reach planned capacities. The primary reason for the delays is believed to be a shortage of funds.

At the inauguration of the plant in 1988, the then President of Brazil's National Nuclear Energy Commission (CNEN), Rex Nazare Alves, said that the facility would produce low-enriched uranium for existing nuclear power and research reactors, and for nuclear submarine reactors. Originally, the centrifuge programme was based at the Institute of Energy and Nuclear Research (IPEN), located on the campus of São Paulo University. It was there, in September 1982, that technicians first succeeded in producing slightly enriched uranium. Two years later, they operated their first cascade, composed of nine machines.

Although the Brazilian Government has released few details about the capabilities of its centrifuges or its capacity to make them, enough public information is available to draw some conclusions.

Centrifuge type

The design of the centrifuge is unknown, although indications are that it is not a very capable machine. At the inauguration ceremony, Alves stated that the rotors of the centrifuges were made out of maraging steel. Brazil is one of the few countries in the world that produce maraging steel and, according to Brazilian press accounts and US Government sources, Brazil could produce enough to supply Aramar. The machine is reported to use magnetic top bearings, spin at 60 000 revolutions per second, and have a lifetime of more than five years.[3] The separative capacity of the machines installed at Aramar is 1.8 SWU per year.

Number of centrifuges

Brazilian news reports and senior Argentine nuclear officials who attended the inauguration ceremonies in 1988 stated that Aramar had about 50 operating centrifuges at this time. There are conflicting reports about the number in operation now. In late 1990, the plant was reported to have 550 new machines, and 48 older units in operation.[4] In mid-1991, about 500 machines were reportedly operating.[5] The plant's first construction phase, apparently not completed at the end of 1991, called for 958 operating centrifuges with a total capacity of about 1700 SWU a year.[6]

A working group appointed by President Collor to develop recommendations about the future direction of the Brazilian nuclear programme stated in a 1990 report that the second construction phase of the Aramar plant should involve an expansion to 100 000 SWU per year by 1996, at an estimated cost of more than $300 million.[7] This new plant would use improved centrifuges with a capacity of 3 SWU per year. The Collor Administration has made no decision about this expansion.

The group's report said that these newer machines have already been developed, that a 10 SWU per year machine is being developed, and that a 25 SWU per year machine is being designed. Some of these machines might use more advanced carbon-fibre rotors, which are capable of much higher speeds than

[3] 'Ultracentrifuges detailed', *O Estado de Sao Paulo*, 18 Dec. 1990 (in Portuguese).

[4] Godoy, R., 'Ultracentrifuges to raise uranium production', *O Estado de Sao Paulo*, 18 Dec. 1990 (in Portuguese).

[5] Casado, J., 'Reaction to accords with IAEA and Argentina: inspection not intrusive', *Gazeta Mercantil*, 30 July 1991 (in Portuguese).

[6] Pinguelli Rosa, L. P., Barros, F. de S. and Barreiros, S. R., 'Nuclear technology in Brazil: retrospective, current situation, and outlook on the nuclear power plan and military/nuclear programme', undated (portions translated by US State Department). Luiz Pinguelli Rosa served as a technical advisor to the Brazilian Congressional Commission investigating the autonomous nuclear programme in 1990 and 1991.

[7] Motta, P., 'Nuclear development recommendations made', *O Globo*, 30 Sep. 1990 (in Portuguese).

those made with maraging steel, resulting in machines with a significantly greater capacity to produce enriched uranium. Carbon-fibre rotors have been under development since at least the inauguration of the Aramar facility, according to Alves.[8]

Enrichment product

At the time of the inauguration, the plant was producing an undisclosed amount of 5 per cent enriched uranium. In mid-1991, it was reported to be producing uranium enriched to a little more than 20 per cent.[9] The amount of enriched uranium produced so far is not publicly known, and the authors do not have sufficient information to estimate it.

Brazil has imported unsafeguarded enriched uranium. About 200 kg of 4.3 per cent enriched uranium were obtained abroad from China or South Africa for a critical facility started in 1984.[10] A South African nuclear official denied that South Africa provided enriched uranium to Brazil.

Production of HEU

Like a gaseous diffusion enrichment plant, a centrifuge plant designed to make up to 20 per cent enriched uranium can be used to produce weapon-grade uranium. But the centrifuge system is more flexible than the diffusion system.

In a gas centrifuge plant designed to produce low-enriched uranium, batch recycling and connecting cascades in series provide an efficient, relatively quick route to weapon-grade uranium. The small amount of uranium found in the cascade and the modular design of most gas centrifuge plants make stretching unnecessary. But by combining a moderate amount of stretching with batch recycling, a plant such as Aramar could produce weapon-grade uranium with one recycle.

When the Aramar plant reaches its planned capacity of 1700 SWU per year, and if Brazil produces an adequate stockpile of 20 per cent enriched uranium, Brazil could then produce 25 kg of weapon-grade uranium, more than enough for a crude nuclear explosive, in about six weeks. If Brazil had to start with natural uranium, it could produce 25 kg of weapon-grade uranium in about three years if all the centrifuges were connected together into one cascade.

Brazil's secret plutonium programme

While the Navy was pursuing the uranium option to the bomb, the Army was developing the plutonium route. In September, 1991 the Army publicly revealed its unsafeguarded nuclear programme, located at the Army Technolog-

---

[8] See also Fantini F. and Costa, R., 'History of nuclear parallel programme surveyed', *Istoe*, 13 Apr. 1988 (in Portuguese).
[9] See Casado (note 5).
[10] See Pinguelli Rosa *et al.* (note 6).

ical Centre (CETEX) in Rio de Janeiro State. It is currently operating a small sub-critical graphite unit containing bars of natural uranium that can only be operated when an external neutron-emitting source is in the reactor.

The Army has been designing a natural uranium, graphite-moderated, air-cooled reactor, called the Experimental Irradiation Reactor. Originally intended to have a power rating of 20 MWth, large enough to make 4 kg of weapon-grade plutonium a year, the reactor has been scaled back to 2 MWth. According to General Nelson de Almeida Querido, of CETEX, the size of the reactor has been reduced to 'make it clear that the Army has no intention of getting plutonium to make nuclear weapons'.[11] The reactor could be finished by the mid-1990s, although the reduction in power has eliminated its plutonium production potential, raising doubts that it will be built.

For several years, Brazil has operated an unsafeguarded laboratory-scale plutonium separation facility at IPEN. The head of IPEN said that this process was demonstrated using plutonium simulators, such as thorium, cerium and gadolinium, since all existing spent fuel in Brazil is safeguarded and subject to foreign control.[12] The confidential 1990 nuclear report to President Collor recommended that the continuation of the laboratory work on plutonium separation at IPEN be followed eventually by a pilot separation plant.

Questions remain about whether uranium targets were irradiated in safeguarded reactors, and processed in the IPEN facility, extracting gram quantities of plutonium. According to a review of the Brazilian nuclear programme prepared by West German intelligence for the Bonn Government, the facility was operated at a throughput of less than 1 gram of plutonium per day.[13] However, a senior Brazilian nuclear official said in an interview in 1988 that the plant would take 'one hundred years to produce a kilogram of plutonium'.

III. South Africa

In July 1991, South Africa became a party to the NPT, citing as a reason the end of the cold war. Foreign Minister R. F. Botha said that South Africa had developed 'the capacity and potential to produce a nuclear explosive device'.[14] He declined to confirm whether South Africa had ever made a nuclear weapon.

South Africa's technical capability to make nuclear weapons has rested on its uranium enrichment programme. South Africa has acknowledged making 45 per cent enriched uranium in an unsafeguarded pilot enrichment plant at Valindaba, near Pretoria. But it has declined to say whether it produced weapon-grade uranium before this plant was closed permanently in February

[11] Motta, P., 'Secretary discusses Army nuclear project details', *O Globo*, 16 Sep. 1990 (in Portuguese).
[12] Zygband, F., 'Nuclear waste reprocessing technology mastered', *O Globo*, 16 Sep. 1990 (in Portuguese).
[13] Hibbs, M., 'Germans say Brazil developing two production reactors', *Nucleonics Week*, 27 July 1989.
[14] Ottaway, D. B., 'South Africa agrees to treaty curbing nuclear weapons', *Washington Post*, 28 June 1991, P. A25.

1990. *Nuclear Fuel* reported in September 1992 that information supplied to the IAEA as part of verifying South Africa's initial inventory declaration indicates that this facility did produce several hundred kilograms of weapon-grade uranium.

Enrichment programme

South Africa began secretly developing a uranium enrichment process in the early 1960s.[15] The existence of this programme was known to only a few people within the government until 1970, when Prime Minister John Vorster announced that South African scientists had developed a unique and economical enrichment process based on the aerodynamic technique.

Because of the high energy costs of this process, however, South Africa also began developing both gas centrifuge and laser enrichment processes. South Africa has investigated the feasibility of building an industrial-scale centrifuge and laser enrichment plant.[16] In August 1991, the gas centrifuge programme was terminated, although the laser programme continues.

Pilot enrichment facility

Following Vorster's announcement in 1970, the government formed the state-controlled Uranium Enrichment Corporation (UCOR) to build the pilot-scale plant, called the Y-Plant, at Valindaba, next to the National Nuclear Research Centre at Pelindaba. The meaning of the name, Valindaba, gives an indication of the secrecy that has surrounded this project. Drawn from the indigenous dialects of South Africa, Valindaba literally means 'the council is closed' or more figuratively 'no talking about this'.[17]

The Y-Plant began commissioning in 1974, and commenced HEU production in January 1978. After overcoming several mechanical and chemical problems, including a halt in production lasting from August 1979 until July 1981, the plant was operated successfully until its closure in 1990.[18] Its capacity is secret, although its nominal capacity is believed to be between 10 000 and 20 000 SWU per year. At this capacity, it could theoretically produce 50–100 kg of weapon-grade uranium per year, starting with natural uranium and using a tails assay of 0.3 per cent. South African scientists responsible for developing the more advanced aerodynamic process, based on the 'helicon' cascade approach, stated in 1976 that the theoretical separative work capacity of a

[15] Newby-Fraser, A. R., *Chain Reaction: Twenty Years of Nuclear Research and Development in South Africa* (The Atomic Energy Board: Pretoria, 1979), p. 95.

[16] Kemp, D. M., Bredell, P. J., Ponelis, A. A. and Ronander, E., 'Uranium enrichment technologies in South Africa', Atomic Energy Corporation of South Africa, Ltd, Paper presented at the International Symposium on Isotope Separation and Chemical Exchange Uranium Enrichment, 29 Oct.–1 Nov. 1990, Tokyo, Japan.

[17] See Newby-Fraser (note 15), p. 104.

[18] See Kemp *et al.* (note 16); and IAEA, *South Africa's Nuclear Capabilities (GC(XXXV)/RES/567)*, IAEA document GC(XXXVI)/1015, Vienna, 4 Sep. 1992.

'prototype module' for the nearby mini-Z plant was 6000 SWU per year.[19] The module consisted of one set of compressors and one set of separation elements.[20] Further developmental work enabled its capacity to be increased to slightly over 10 000 SWU per year.[21]

This plant has often been confused with the Y-Plant, but is unrelated. The prototype module was closed about five years ago after successfully demonstrating the feasibility of the helicon concept, and produced very little enriched uranium.

In 1981, South Africa announced that the Y-Plant had succeeded in producing a limited amount of 45 per cent enriched uranium for use in the Safari I research reactor.[22] Until now, South Africa has declined to acknowledge production of weapon-grade uranium or the existence of a nuclear weapon programme.

The semi-commercial-size enrichment plant

South Africa has built a much larger enrichment plant at Valindaba based on the same separation techniques as used in the Y-Plant, but using the quite different helicon approach. Construction of this larger plant began in 1979, and was completed at the end of 1986. Because of problems resulting from insufficient prototype experience, production did not begin until 1988.[23] As of the end of 1991, the plant could operate at its optimum production of 300 000 SWU per year. This plant is intended to produce 3.25 per cent uranium for the twin Koeberg power reactors, which require about two-thirds of its optimum annual production. Any spare separative capacity could be sold on the world market.[24]

The plant is widely reported not to be designed to produce highly enriched uranium. It could theoretically produce highly enriched uranium through either batch recycling or 'stretching' the cascade, however, but use of the plant to produce highly enriched uranium is impractical and could create criticality problems in the plant. The cascades in the plant are capable of achieving an enrichment level of 3.25 per cent only after three successive batching operations, using a 0.3 per cent tails assay.[25]

---

[19] Benedict, M., Pigford, T. and Levi, H., *Nuclear Chemical Engineering* (McGraw Hill: New York, 1981), p. 889. See also Kovan, D., 'Enrichment plants—a survey of major new uranium enriching projects', *Nuclear Engineering International*, Nov. 1976, citing Manson Benedict at an Atomic Industrial Forum conference in June 1976.

[20] See Newby-Fraser (note 15), p. 111.

[21] See Newby-Fraser (note 15), p. 111.

[22] Abendroth, K., 'Enriched uranium produced for research reactor', *The Citizen* (Johannesburg), 29 Apr. 1981, p. 3.

[23] See Kemp *et al.* (note 16).

[24] Jones, J., 'South Africa enrichment plant now commercial, AEC head says', *Nucleonics Week*, 23 Jan. 1992. The production of enriched uranium for the Koeberg reactors requires about 200 000 SWU per year. The remaining 100 000 SWU per year would theoretically be available for sale.

[25] See Kemp *et al.* (note 16).

Gas centrifuge and laser enrichment

Because of the high energy costs of the aerodynamic enrichment process, South Africa decided to develop both gas centrifuge and molecular laser enrichment techniques.

South Africa began centrifuge research in the 1970s.[26] Its centrifuges are based on European-style centrifuges, since South African enrichment experts state that a 'large amount of knowledge of these designs was in the public domain'.[27] It recently decided to discontinue its gas centrifuge programme, because of the difficulty of ever developing a centrifuge plant competitive with existing plants in Europe.

South Africa began its efforts in molecular laser isotope separation in 1983.[28] The programme is in an early stage of development, although the first demonstration module is scheduled to operate in 1993, followed by a pilot plant in 1997.

Inventory of highly enriched uranium

South Africa has declared its inventory of highly enriched uranium to the IAEA, but has not released the information to the public. Despite the lack of this information, we estimate a potential inventory of weapon-grade uranium.

If the Valindaba plant had a maximum capacity of 10 000–20 000 SWU per year from the beginning of 1978 to August 1979, and again from July 1981 until its shut-down in early 1990, it could have produced up to 500–1000 kg of weapon-grade uranium, assuming that natural uranium was used as feed and the tails assay was 0.3 per cent. Several factors, however, would reduce this estimate. Since the plant is known to have had start-up problems, and to have encountered operational difficulties, an average annual capacity of 65 per cent is assumed. This assumption would lower the above estimate to about 325–650 kg of weapon-grade uranium.

Some fraction of Valindaba's capacity was dedicated to producing 45 per cent enriched uranium for the Safari research reactor. Although designed to operate at 20 MWth, this reactor has operated at only 5 MWth since the 1970s, according to a South African nuclear official at the Safari reactor interviewed

---

[26] See Kemp *et al.* (note 16). South Africa developed both maraging steel and carbon-fibre/resin centrifuges. Although the latter are more difficult to develop, particularly given South Africa's indigenous capabilities, a plant using carbon fibre centrifuges would be significantly less expensive than one using maraging steel machines. The programme aimed to develop a maraging steel machine with a rotor 2 m long and 0.145 m in diameter spinning at 450 m/s with a separative capacity of 10 SWU per year. It hoped to develop a carbon-fibre design that would have had a rotor 3 m in length and 0.175 m in diameter spinning at 550 m/s with a capacity of 30 SWU per year. The authors of this report write that they had a laboratory with 16 stands, which test various mechanical properties of the centrifuges, and another rig, which used uranium hexafluoride to measure the separative capacity of a machine. Single carbon-fibre machines were being developed, and cascade development would have begun only after additional single machine tests of bearing lifetimes and of longer rotors. The report does not discuss the maraging steel centrifuges.

[27] See Kemp *et al.* (note 16).

[28] See Kemp *et al.* (note 16)

by one of the authors in early 1992. Each year, this reactor requires roughly 10 kg of 45 per cent enriched uranium, requiring about 1000 SWU per year. If the pilot plant produced enough enriched uranium to fuel this reactor up to the end of 2005, when the reactor would be 40 years old, it would have used about 25 000 SWU. This reduces our estimate by about 125 kg of weapon-grade uranium.

South Africa could have produced a much larger amount of weapon-grade uranium if it had started with a supply of low-enriched uranium as feedstock. Material South Africa reportedly obtained from the People's Republic of China, nominally for use in the Koeberg reactors, is a possible source of low-enriched uranium.[29] *Nucleonics Week* reported that a West German middleman arranged for the export from China to South Africa of 30 tonnes of 3 per cent enriched uranium and 30 tonnes of 2.7 per cent enriched uranium in the form of uranium hexafluoride.[30] If used as feedstock, this would be enough low-enriched uranium to produce about 1300 kg of weapon-grade uranium. South Africa, however, is unlikely to have used this material as feedstock in the enrichment plant, since it stockpiled this uranium as a reserve for the Koeberg reactors in case of delays in starting the semi-commercial enrichment plant.

Assuming that low-enriched uranium was not used as feedstock, and that the semi-commercial plant has not produced any highly enriched uranium, the authors estimate that South Africa produced a potential inventory of about 200–525 kg of weapon-grade uranium.

Without more information, this estimate is highly uncertain. Although the intelligence agencies of the United States and the former Soviet Union should have more highly developed estimates, a significant degree of uncertainty is likely to surround their estimates as well. This creates a serious challenge to the IAEA to verify independently South Africa's declaration of its enriched uranium inventory. South Africa has given the IAEA extensive historical production information on its enrichment plants, but this information is insufficient to eliminate all discrepancies between South Africa's declaration and the IAEA's estimate of that declaration. The IAEA, however, found no evidence that South Africa's declaration is incomplete.

Plutonium separation

South Africa began operating a hot cell facility at Pelindaba in 1987 to examine spent fuel from the Koeberg reactor, although it is not known whether this facility can also separate small amounts of plutonium.[31]

---

[29] See, for example, The Washington Post Wire Service, 'Chinese uranium reportedly sold for South African nuclear programme', *Atlanta Journal*, 19 Nov. 1981, p. 23-A.

[30] Hibbs, M., 'British report details heavy water uranium trades by Hempel firms', *Nuclear Fuel*, 25 July 1988.

[31] South African Embassy, Washington, DC, 24 July 1987; and 'South Africa's aim is "self-sufficiency" as AEC studies casks and reprocessing', *Nuclear Fuel*, 24 Mar. 1986.

South Africa has stated that it would like to reprocess spent power reactor fuel in the next century.[32] However, under a bilateral agreement with France, which supplied the Koeberg reactors, South Africa is barred from reprocessing on its territory Koeberg spent fuel containing uranium supplied by France. Furthermore, any plutonium separated from such fuel outside of the country may not be returned to South Africa.

South Africa is also considering interim and final disposal of Koeberg spent fuel in a repository at Vaalputs in the south-western part of the country. This could be an alternative to reprocessing.

IV. Taiwan

Since 1985, Taiwan has been obliged to send spent fuel containing high-quality plutonium to the United States to reduce the risk that Taiwan might separate the plutonium for use in nuclear weapons. The fuel was discharged from the 40-MWth Taiwan Research Reactor (TRR). As of the end of 1991, almost all of the fuel had been sent, but an environmental lawsuit in the USA has delayed the last shipment pending the completion of an environmental assessment accept-able to the court.

The TRR, supplied by Canada in 1969, has the same design as the Cirus reactor in India. Since the plutonium is of high quality, the US Government was worried that Taiwan might separate it for weapon use.

Such concerns were evidently legitimate. In 1987 the USA learned that Taiwan was secretly building an installation capable of extracting the pluto-nium from TRR spent fuel.[33] In response to US pressure, Taiwan stopped work on this installation and permanently shut down the TRR.

This is not the first time that the United States has been concerned about Taiwan's intentions. According to a 1974 CIA assessment, Taiwan conducted 'its small nuclear programme with a weapon option clearly in mind'. The TRR was widely viewed as the centre-piece of that programme, and in 1977 the United States pressured Taiwan to close a hot cell it was building that could have separated plutonium from TRR fuel.

The TRR operated for about 14 years, and during this time discharged about 1600 fuel rods. Each rod contained an estimated 70 grams of plutonium, with more than 90 per cent of it being ^{239}Pu.[34] Thus, in total the reactor produced an estimated 110 kg of high-quality plutonium. If 5–6 kg are assumed per atomic bomb, Taiwan had enough plutonium for about 20 weapons.

Despite the merits of these spent fuel shipments for non-proliferation, they conflicted with safety and environmental concerns within the United States.

[32] See *Nuclear Fuel* (note 31).

[33] Engelbery, S. and Gordon, M. R., 'Taipei halts work on secret plant to make nuclear bomb ingredi-ent', *New York Times*, 23 Mar. 1988.

[34] This value is derived from US Department of Energy, 'Environmental assessment on shipment of Taiwanese research reactor spent fuel (Phase II)', DOE/EA-0363, June 1988. Table D.1 lists the radio-active inventory of the fuel rods, where a burn-up of 1600 MWth-d per tonne of heavy metal is assumed.

Originally, the fuel was to be sent by ship to the west coast of the United States, and then by road to the Savannah River Site in South Carolina for processing, but intense local opposition forced the US Government to send them directly to the east coast and then to the Savannah River Site.[35] Since the uranium which gave rise to the spent fuel was not of US origin, opponents of the shipments worried that this could open the way to the United States becoming a dumping ground for foreign spent fuel. They also objected to the fuel being reprocessed in military facilities, although the US Government said that the Taiwanese plutonium would not be used in weapons.

Recently, the last shipment was stopped by a Federal court that ruled that the government's environmental assessment of the consequences of an accident was inadequate. It is unclear when this shipment will occur. However, most of the plutonium-bearing spent fuel had already left Taiwan.

[35] Cipriano, R., 'Nuclear fuel rods re-routed to Virginia', *Los Angeles Times*, 11 July 1986.

Part V
Conclusions

Two chapters bring the book to a conclusion. Chapter 12 presents an overview of the results of chapters 3–11. Section I summarizes our findings on the sizes, locations and forms of plutonium and HEU inventories at the end of 1990, the most recent year for which a complete set of data can be compiled. It also assesses the amounts of material that are presently under international safeguards, and compares the aggregate statistics published by Euratom and the International Atomic Energy Agency with our own estimates. Section II considers possible trends in plutonium and HEU inventories over the next two decades.

Chapter 13 contains a brief discussion of the two main policy issues that we see emerging from this book. The first concerns the poor quality of public information about the majority of plutonium and HEU that exists in the world, and the resulting difficulty of establishing confidence that they are held under proper control. In particular, it seems unacceptable that error margins of tens of tonnes are attached to inventories in the nuclear weapon states, while inventories elsewhere have to be accounted for with error margins of a few kilograms. The second concerns the need to develop strategies for coping with the large amounts of plutonium and HEU that will emerge from civil spent fuels and from nuclear weapons in years ahead. While the HEU can all be used as a nuclear fuel, a substantial proportion of the plutonium may have to be treated as waste.

12. Overview of present and future stocks of plutonium and highly enriched uranium

Throughout this book estimates have been presented with approximate error margins. This chapter presents in most places central estimates, without error margins, primarily for reasons of clarity. The reader is encouraged to consult earlier chapters for assessments of relevant levels of uncertainty.

All plutonium figures cited in this chapter are quantities of 'total plutonium' (Pu_{tot}). This follows the convention adopted by the safeguards agencies.[1]

I. Inventories at the end of 1990

Aggregate totals

The central estimates of the world inventories of plutonium and HEU at the end of 1990, rounded to two significant figures, are:

For plutonium: 910 tonnes
For HEU: 1300 tonnes

Precise error margins cannot be attached to these figures. The authors feel comfortable with error margins of plus or minus 15 per cent for plutonium, and plus or minus 30 per cent for HEU. These margins are strongly influenced by uncertainties over the sizes of military inventories in the nuclear weapon states, and in the former Soviet Union in particular.

It should be noted that the above figures exclude (*a*) the amounts of HEU in the naval fuel cycles and in production reactor fuel cycles in the NWS; (*b*) the amounts of plutonium in operating power reactors (inclusion of *a* and *b* would add approximately 90–95 tonnes—in power reactors—to the stock of plutonium, and very approximately 100–200 tonnes to the stock of highly enriched uranium[2]); (*c*) amounts of HEU in NWS breeder reactors and research facilities; (*d*) amounts of HEU discharged from civilian research reactors in the USA,

---

[1] See chapter 2 for an explanation of the differences between Pu_{tot} and Pu_{fiss}.

[2] The latter range is based on US figures. About 90 tonnes of HEU are dedicated to the US naval programme (see chapter 4). Approximately 25 t of HEU (40–60% enriched) are also tied up in the fuel cycle of the Savannah River production reactors. Some small fraction of this material is HEU recovered from naval fuel, but its effect on this estimate is negligible.

We do not know how many CIS naval and production reactors use HEU fuel. It is likely that these fuel cycles would contain less HEU than their US equivalents. As a result, we estimate that the amount dedicated to CIS naval and production reactor fuel cycles has an upper bound of 100 tonnes. The amounts in the fuel cycles of the other NWS would not affect these estimates significantly.

China and the CIS and returned to the military programme: and (*e*) any HEU fuel returned to military programmes after being discharged from foreign research reactors. Inclusion of (*d*) and (*e*) would add to the stock of highly enriched uranium approximately 30 tonnes (enriched to about 70 per cent) (see chapter 8). The reason for excluding these quantities is that it is very difficult to assess the fate of this material. In the case of the USA, the HEU recovered from spent fuel from domestic reactors and returned from foreign reactors is assigned to the fuel cycle of the Savannah River military production reactors. Thus its inclusion would result in double counting. In addition, this material has undergone additional fissioning, reducing the amount remaining. Amounts of plutonium contained in MOX fuels inserted in fast and thermal reactors are, however, included in the 910 tonnes total.

In general, public knowledge of HEU inventories is less developed than that of plutonium inventories.

Civil and military inventories

The above inventories are broken down in table 12.1 into their civil and military components. Again these are central estimates. This table shows that the great majority of HEU is contained in military inventories, which consist largely of weapon-grade uranium. The stock of weapon-grade plutonium is also large, but it is now exceeded by the amounts of fuel- and reactor-grade plutonium existing in the civil sector.

These figures relate to the end of 1990. The inventory of weapon-grade HEU may have diminished slightly since then. Production has ended in the USA and the former Soviet Union, and if production continues in China and France, the amounts will be relatively small. During 1991 and 1992, there may have been modest draw-downs on these HEU inventories, mainly for submarine reactors.

The stock of military plutonium will also have changed little in 1991 and 1992. Some production still continues in Russia. Elsewhere the supply of plutonium for military purposes has either ceased, or is occurring at a relatively low rate (e.g., in France, Israel and perhaps India).

In contrast, the increase in the civil inventory of plutonium will have been substantial since 1990. The world stock of power reactors is discharging approximately 9300 tonnes of spent fuel each year, containing some 62 tonnes of plutonium. The combined military and civil inventory at the end of 1991 was therefore around 970 tonnes. If our estimates are correct, the 1000-tonne threshold was crossed early in 1992. Had the amounts of plutonium still in the cores of operating reactors been included in these figures, this milestone would have been passed in 1990.

Table 12.1. Central estimates for civil and military inventories of plutonium and HEU, end of 1990

Figures are in tonnes and do not include material in reactor cores.[a]

	Civil inventory	Military inventory	Total
Plutonium	654	257	911
HEU	20[b]	1 310	1 330

[a] Totals do include material recycled in fast reactor cores.

[b] This includes HEU in civilian programmes in both the non-nuclear and nuclear weapon states.

Table 12.2. Central estimates for inventories of plutonium and HEU by type, end of 1990

	Tonnes
Civil plutonium	
In spent reactor fuel	532
Separated in store	72
In fast reactor fuel cycle	37
In thermal MOX fuel cycle	13
Civil HEU	
In research reactor fuel	20
Military plutonium	
In warheads	178
Weapon-grade outside warheads	56
Fuel and reactor-grade in store	23
Military HEU	
In warheads	810
Outside warheads (not including submarines)	500

II. Types of inventory

The quantities in table 12.1 can be broken down further, according to the forms and mediums in which they are held. The results are shown in tables 12.2 and 12.3. The majority of plutonium in the civil inventory is contained in spent fuel stores. Two-thirds of the plutonium that has been separated also remains in store. Of this stock, around 25 tonnes is in Russia, most of the remainder being held in Western Europe (including 34 tonnes in the UK). The 37 tonnes of plutonium in the fast reactor fuel cycle includes material in reactor cores, spent fuel ponds, and being reprocessed or fabricated into fresh fuel elements. The same applies to the estimated 13 tonnes in the thermal mixed-oxide (MOX) fuel cycle.

Table 12.3. Central estimates of highly enriched uranium in the nuclear weapon states, end of 1990

Country	Tonnes of HEU
China	15
France	15
UK	10
CIS	720
USA	550
Total	**1 310**

We have divided the inventories of military plutonium and HEU into two main stocks: those inside and those outside weapons. These allocations should be regarded as very rough guides to the actual amounts. They are based on the assumption that there are 52 000 warheads extant—32 000 in the CIS, 19 000 in the USA, and 1000 in China, France and the UK taken together (rounded to the nearest thousand). Using assumptions set out in section V below about the average amounts of plutonium and HEU in warheads, we estimate that this arsenal contains 178 tonnes of plutonium and 810 tonnes of HEU. It should be noted that the amounts of military HEU *outside* warheads are substantial, and larger than the amounts of plutonium.

III. Location of inventories by NPT status

No more than 15 tonnes, or 1 per cent of the world stock, of HEU is located outside the NWS. Furthermore, much of this HEU was produced by the USA. Our central estimate of 1310 tonnes of HEU in the NWS is made up as shown in table 12.3.

These figures carry generous error margins. But they show that over 95 per cent of the stock held by the NWS is located in the USA and the CIS. The production of weapon-grade uranium has largely been their preserve (most of the British stock has also been supplied by the USA).

Prior to South Africa's accession to the NPT, no HEU had been produced by NNWS parties to the Treaty. The authors have estimated that South Africa could have acquired 200–525 kg of weapon-grade uranium by the end of 1991. Among the non-NPT threshold countries, only Pakistan is believed to have produced quantities of weapon-grade uranium that give it a substantial weapon capability (the estimate is that it could have produced 130–220 kg by the end of 1991). Argentina, Brazil, India, Iraq and Israel each have enrichment capabilities at varying stages of development, but there is no evidence that they have produced HEU in the amounts required for nuclear weapons.

Inventories of plutonium are more dispersed across NWS and NNWS, and NPT and non-NPT countries, as is shown in table 12.4. Nevertheless, most of

Table 12.4. Central estimates for plutonium inventories, by NPT status, end of 1990
Figures are in tonnes.[a]

Inventory	NPT signatory states		Non-signatory states[b]	Total
	NWS	NNWS		
In warheads	178	–	0.3	178
In other military stockpiles	79	–·	0.5	79
In spent reactor fuel	$\left\{ \begin{array}{c} 297^c \\ 366^d \end{array} \right.$	$\left. \begin{array}{c} 218^c \\ 148^e \end{array} \right\}$	17	532
Separated civil plutonium	65[f]	7[g]	0.3	72
Recycled civil plutonium	31[h]	19[i]	–	50
Total (by ownership)	**649**	**244**	**18**	**911**
Total (by location)	**719**	**174**	**18**	**911**

[a] Error margins for military stocks inside and outside warheads are more substantial than elsewhere in the table. The most accurate figures are those for stored spent fuel.

[b] Includes Taiwan.

[c] The figure in the first column includes an estimated 6 t of plutonium contained in Eastern European and Finnish fuel sent to Chelyabinsk and Krasnoyarsk under 'take-back' arrangements. The figure in the second column excludes this amount.

[d] Includes the estimated 70 t of foreign plutonium held in store in France and the UK, in spent fuel or as separated plutonium.

[e] Indicates the amount of irradiated fuel left in NNWS after transfers to France and the UK.

[f] Includes an estimated 2 t of plutonium separated at Sellafield and La Hague but not yet returned to owners in NNWS.

[g] Comprises 28.4 t of plutonium separated from NNWS fuel, minus 19 t of recycled material and 2 t held in store in Britain and France.

[h] Comprises 29.8 t of civil plutonium recycled in fast reactors and other R&D facilities (5 t in the UK; 17 t in France; 0.5 t in the CIS; and 6.7 t in the USA, including material imported from the UK), and 1.1 t recycled in thermal reactors in France. The 17 t recycled in French fast reactors includes 4.5 t of foreign plutonium in Superphénix fuel.

[i] Comprises 4.5 t and 3.2 t recycled in Japanese and German fast reactors, respectively, and 11.4 t recycled in German, Belgian and Swiss thermal reactors.

the materials are still held in the NWS. Some 70 per cent of the plutonium discharged from reactors is the property of the NWS. In fact, a considerably larger proportion (close to 80 per cent) of plutonium is *located* in the NWS, since some 70 tonnes of plutonium contained in NNWS spent fuels have been transferred to France and the UK for reprocessing where it will remain until the plutonium is separated and returned. We estimate that another 23 tonnes of plutonium had already been separated from NNWS fuel in France and the UK at the end of 1990. A small proportion of this (perhaps 2 tonnes) was still awaiting return to its country of origin. In addition, some 5.5 tonnes of plutonium had been separated by NNWS parties to the NPT.

The estimated 257 tonnes of plutonium in the military stockpiles of the NWS is divided as shown in table 12.5.

Table 12.5. Estimated plutonium in the military stockpiles of nuclear weapon states, end of 1990

Country	Tonnes of plutonium
CIS	125
USA	112
UK	11
France	6.0
China	2.5
Total	**257**

Table 12.6. Central estimate of plutonium in India and Israel, end of 1991

Country	Kilograms of plutonium
Israel	330
India	290

The US and British inventories include 15 and 7.6 tonnes of fuel-grade plutonium, respectively. It is not known whether the CIS, France and China have stocks of fuel-grade material in their military inventories. The error margins are also substantial here, particularly for the CIS and China, although they are probably less than in the case of HEU.

The amounts of plutonium separated by non-NPT countries and held outside safeguards are considerably larger than the corresponding amounts of HEU. The main inventories belong to Israel and India. Smaller amounts of plutonium may also have been separated by Argentina and Pakistan. Our central estimates for Israel and India at the end of 1991 are rather similar (table 12.6).

A substantial proportion of the Israeli inventory is probably contained in nuclear weapons. If Israel has manufactured this plutonium into weapon cores, each containing 5 kg of plutonium on average, its nuclear stockpile would be sufficient to manufacture around 65 weapons.

It is worth noting here that none of the threshold countries has succeeded in acquiring sizeable inventories of both HEU and plutonium. Together with the lack of weapon testing, this probably constrains the types of nuclear warhead that any one of them can construct at present.

IV. Material under international safeguards

The primary function of international safeguards is to detect diversions of materials from civil to military use (the IAEA), or from the purpose declared by owners of materials (Euratom). Under the NPT, safeguards are mandatory for NNWS but not for NWS parties. The NWS have individually entered into 'voluntary offer' agreements with the IAEA, whereby they inform the Agency of the facilities they are prepared to submit to safeguards.[3] The Agency can in turn choose ('designate') which facilities it will safeguard at a given time. In contrast, Euratom *has* to safeguard declared civil material in all member states, including the NWS (France and the UK).

The result is that the IAEA safeguards all known inventories of plutonium and HEU located in the NNWS which are parties to the NPT. However, it safeguards only a small proportion of the material held by the NWS, and the majority of that is material from Japan and European NNWS temporarily located in France and the UK under reprocessing arrangements. No HEU or separated plutonium in the USA is safeguarded by the IAEA, and only small amounts, contained in spent fuel from one power reactor and one research reactor, are safeguarded in the CIS (the amounts will increase significantly when Kazakhstan and Ukraine join the NPT regime). In contrast, substantial proportions of the plutonium inventories of France and the UK are safeguarded by Euratom (their HEU inventories are largely assigned to weapons and thus outside Euratom safeguards).

The IAEA also endeavours to safeguard materials possessed by threshold countries not party to the NPT. However, a large percentage of their separated plutonium and HEU remains unsafeguarded, especially in Israel, India and Pakistan.

Each year, the IAEA and Euratom publish aggregate figures for the plutonium and HEU which they safeguard. They are barred by rules of confidentiality from disclosing the amounts safeguarded in individual countries. At the end of 1990, the IAEA and Euratom safeguarded around 12 tonnes and 13 tonnes of HEU respectively (the quantities overlap since they both inspect material in the European NNWS).[4] Less than 1 per cent of the world stock of HEU is therefore under IAEA safeguards. Furthermore, the IAEA figure is not adjusted to take account of burn-up and is thus higher than the actual amounts contained in the safeguarded fuel.

Much larger quantities of plutonium are under international safeguards. Euratom safeguarded 203 tonnes of plutonium in the European Community (EC) at the end of 1990, of which 'approximately 30 per cent' was in separated form.[5] These figures are in close accord with our own estimates of the amounts of civil plutonium produced and separated by EC countries. Those estimates are that 203.5 tonnes of plutonium had been produced by the end of 1990, and

---

[3] The model NPT safeguards agreement is set out in an IAEA information circular (INFCIRC)—IAEA document INFCIRC/153 . Materials in NNWS that are not parties to the NPT are safeguarded under the earlier INFCIRC/66/Rev. 2. The INFCIRC numbers of the voluntary offer agreements are 263 (UK), 288 (USA), 290 (France), 327 (USSR) and 369 (China).

[4] Euratom safeguards a larger amount of HEU because unlike the IAEA it safeguards all civil material in France and the UK. See Euratom, *Report on Operation of Euratom Safeguards*, SEC/92/80 Final (Euratom: Luxembourg, 24 Jan. 1992).

[5] Euratom (note 4).

Table 12.7. Approximate quantities of material subject to IAEA safeguards, end of 1990

Figures are in tonnes.[a]

	NNWS			
	NPT	Non-NPT	NWS	Total
Plutonium contained in irradiated fuel[a]	⎰ 212.4	22.2	75[b]	309.6
	⎱ 154.6	16.2	74.5	245.3
Separated plutonium outside reactors	8.5	–	11.6	20.1
Recycled plutonium in fuel elements in reactor cores	1.8	–	–	1.8

[a] Figures in the lower row are net totals after the IAEA's estimated quantities of plutonium in reactor cores (64.3 t of plutonium) are subtracted on a *pro-rata* basis from the totals for NPT and non-NPT NNWS given in the upper row. This material is not reported to the IAEA under agreed reporting procedures. A total of 0.5 t has also been subtracted from the quantity of safe-guarded irradiated fuel in the NWS to cover the inventory of plutonium in the Novovoronezh reactor cores.

[b] This is assumed to include the 11.6 t listed as separated plutonium (see text).

Sources: International Atomic Energy Agency, *The Annual Report for 1990* (IAEA: Vienna, 1991), p. 155.

87 tonnes had been separated, of which around 36 tonnes had subsequently been recycled in fast or thermal reactors.[6]

The plutonium figures published by the IAEA can also be reconciled with our estimates. Table 12.7 shows the estimated amounts of material under safe-guards that were declared in the Agency's 1990 Annual Report. Unlike Eura-tom's figures, the IAEA includes its own estimate of the quantities of pluto-nium still held in reactor cores under the heading 'plutonium contained in irradiated fuel'. After this quantity still in cores is subtracted, the numbers (in the second row in table 12.7) can be compared with our estimates of inventories in the three categories of countries in table 12.4.

The following comments should be made about the correspondence between our figures and those of the IAEA:

1. The main point at which IAEA figures can be used to corroborate our estimate is the entry for plutonium contained in discharged irradiated fuel in NNWS parties to the NPT. The IAEA figure is 154.6 tonnes. Our estimate for the inventory of plutonium in spent fuel is 148 tonnes (see table 12.4). How-ever, this excludes plutonium contained in spent MOX fuel which has been counted under the 'recycled civil plutonium' heading in table 12.4. The authors estimate that 5–6 tonnes of recycled plutonium may have been discharged from reactors, mainly in the FRG and Switzerland, by the end of 1990. This would bring the estimate of plutonium in discharged irradiated fuel up to 153–54 tonnes.

[6] Our estimated quantities of discharged plutonium are (in tonnes): Belgium 11.5; France 78.2; the FRG 38.8; Italy 5.6; Netherlands 2.4: Spain 13.9; the UK 53 (see table 5.2). This gives a total of 203.5 t.

2. The 75 tonnes of plutonium in irradiated fuel in the NWS column in table 12.7 is predominantly NNWS spent fuel held in the spent fuel ponds at La Hague (France) and Sellafield (UK). The only material belonging to the NWS in this category that is under IAEA safeguards is spent fuel from the Novo-voronezh power reactors in Russia, which could be around 4 tonnes. This conforms with the assumption made in table 12.4 (note *d*) that around 70 tonnes of plutonium in foreign spent fuel is held at La Hague and Sellafield.

3. The 11.6 tonnes of separated plutonium safeguarded by the IAEA in the NWS can only be that held in store 9 at Sellafield, since it is the only plutonium store on the facilities lists presented to the IAEA by the NWS. Specifically, this 11.6 tonnes is a quantity of plutonium equivalent to the amount contained in the Japanese inventory of spent oxide fuel held in Sellafield storage ponds.[7] Note that these 11.6 tonnes are also presumably included in the 75 tonnes listed as contained in irradiated fuel that is safeguarded in the NWS, which therefore represented the total inventory of safeguarded plutonium in the NWS.

4. The IAEA's figure for plutonium contained in discharged fuel in non-NPT states is similar to ours:16.2 tonnes compared with 15.9 tonnes.[8]

Overall, at the end of 1990 the IAEA safeguarded 255.6 (245.3 plus 8.5 plus 1.8) tonnes of plutonium discharged from power reactors. This represented around 28 per cent of the world inventory of plutonium, and around 39 per cent of the world inventory of civil plutonium.

V. Possible future trends in plutonium and HEU inventories

It appears that the production of weapon-grade plutonium and HEU in the NWS is approaching its end. The US and Russian governments have announced that supplies of HEU have been cut off, their present stocks being more than adequate to meet all conceivable future military needs. Furthermore, while plutonium will continue to be produced in civil reactors, it seems likely that the production of weapon-grade plutonium is reaching an upper limit, even though three military production reactors still operated in Russia in early 1993. All production reactors in the USA have been shut down and a moratorium on new plutonium production for weapons was announced by President Bush in July 1992.

Efforts to produce weapon-grade materials continue in a few threshold countries. Even there, cut-offs in production may be approaching in some

---

[7] Under an INFCIRC/66-type agreement which still stands (INCIRC/125), the UK is obliged to submit Japan's inventory of plutonium located in the UK to IAEA safeguards. As a facility attachment for the THORP spent fuel ponds in which most of this plutonium is held has not yet been agreed with the Agency (in late 1992), an amount of separated civil plutonium from the UK's own inventory equivalent to that contained in Japan's spent fuel at the THORP plant has been brought under IAEA safeguards since the Japanese oxide fuel began arriving in Britain. As all the foreign spent fuel in the THORP ponds is, however, subject to 'ad hoc' inspections by the IAEA, it is presumably included in the 75 t in the row above in table 12.7.

[8] Our 15.9 t is made up from an estimated 2.2 t of safeguarded plutonium in India, 4.1 t in Argentina, 0.2 t in Brazil, 0.3 t in Pakistan, 1.5 t in South Africa and 7.6 t in Taiwan.

cases. Pakistan has indicated that it has halted HEU production, and Israel is under pressure to cease producing plutonium. Even if the grounds for optimism are misplaced, the amounts produced by threshold countries are likely to remain relatively small. They may have great political and security ramifications, but they are unlikely to have a substantial effect on the size and distribution of world inventories. Many uncertainties surround production rates in these threshold countries, so the authors have not attempted to draw up long-term forecasts. By 1995, India could have produced about 420 kg of weapon-grade plutonium, and Israel could have increased its inventory to almost 375 kg. If Pakistan resumed the production of weapon-grade uranium, it could have produced some 475 kg of this material by 1995.

The major changes will occur in two other contexts. One involves the growth of plutonium inventories because of the irradiation of uranium fuels in nuclear power stations, and the degree to which the resulting spent fuels are reprocessed and separated plutonium is recycled. The other involves the extraction of plutonium and HEU from dismantled nuclear weapons as the NWS implement deep arms reductions. The amounts of material that may come from these two sources are considered next.

Plutonium released in the civilian fuel cycle

Barring catastrophic accidents, the amount of plutonium that will be produced in operating power reactors over the next two decades is reasonably predictable. The capacity of reactors coming into operation (e.g., in Japan) may be roughly offset by the capacity of reactors being shut down (e.g., in the CIS and the UK). The major changes will occur after 2010 when the many reactors which came into operation in the 1970s and early 1980s will approach the end of their lifetimes. Generating capacity thereafter will depend on whether new investment programmes can be launched in the late 1990s and in the early part of the next century.

In relation to plutonium arisings, a significant uncertainty concerns the burn-ups achieved in uranium fuels (see chapters 2 and 5). Table 12.8 gives some projections of spent fuel and plutonium arisings, based on relatively conservative assumptions about nuclear capacity and about the increases in burn-up discussed in chapter 5.

The maximum conceivable rate of reprocessing over the next decade is 3690 tonnes of spent fuel per year (see table 6.11). With regard to oxide fuels, this comprises 650 tonnes per year in the UK, 1600 tonnes per year in France, 250 tonnes per year in Russia, and 90 tonnes per year in Japan. For magnox fuels, there is additional capacity to reprocess 1100 tonnes per year in France and the UK.

In the decade 2001–10, the maximum rate of oxide fuel reprocessing could increase to 3100 tonnes per year if Japan's plant at Rokkasho-mura comes into operation. This also assumes that Russia's RT-1 reprocessing plant at Chelya-

Table 12.8. Projection of cumulative spent-fuel discharges and plutonium arisings, ends of 1990, 2000 and 2010

Figures are in tonnes.

	End of 1990	End of 2000	End of 2010
Spent fuel	114 800	222 000	320 000
Plutonium discharged	653	1390	2100
Plutonium separation	122	< 310	< 550

binsk ceases operation by the end of this century. There has recently been talk of completing the Krasnoyarsk reprocessing plant during the 1990s, but this may be considered unlikely in view of economic conditions in Russia and the lack of need for more separated plutonium. The maximum rate for magnox fuel reprocessing would be 850 tonnes per year, given that no reprocessing of magnox fuels in France is anticipated after 2000 (its Magnox reactors all having been shut down by 1992).

Although these reprocessing rates represent the industry's present plans, they can safely be regarded as upper bounds. No other industrial countries show signs of wishing to establish substantial reprocessing capacities, and the facilities in France and the UK are unlikely to be expanded.

In view of the mounting political and economic concerns over the future of reprocessing, two other scenarios are considered in chapter 6. One is for a phase-out of reprocessing in the early to mid-1990s, total reliance being placed thereafter on a policy of spent-fuel storage (as is now practised in Canada and the USA, among other countries) with gradual draw-down of existing stocks of plutonium. The exception to this would be the reprocessing of magnox fuel for safety and environmental reasons. The other scenario envisages a 'stretch-out' of this decade's reprocessing contracts. In effect, business for 10 years would be extended over 20 years.

The results are shown in table 12.9. As would be expected, the range of possible outcomes is enormous (the continuing plutonium separation under the phase-out scenario is due to the survival of magnox reprocessing). The table shows that less than one-third of plutonium arisings from power reactors would be separated even if present reprocessing plans were implemented. Nevertheless, the quantities separated would still be very substantial by historical standards. These plans would result in an almost fivefold increase in the amount of plutonium separated from power reactor fuel (up from 122 tonnes to 546 tonnes in 2010) early in the next century. If current military inventories are included, and it is assumed that these are not expanded, then the increase will be of the order of 200 per cent over this period (up from 380 tonnes to 800 tonnes). Just as one source of large-scale plutonium separation—the US and Soviet military programmes—appears to have exhausted itself, another may therefore be opening up.

Table 12.9. Scenarios for civil plutonium separation and use up to 1990, in 1991–2000 and in 2001–10

Figures are in tonnes.

	Up to 1990	1991–2000	2001–10	Total to 2010 (Cumulative)
Plutonium arisings				
In spent fuel	654	737	710	2 100
Plutonium separated				
Phase-out	(122)	43	–	165
Stretch-out	(122)	122	126	370
Maximum feasible (current plans)	(122)	188	236	546
Plutonium consumed				
Maximum recycle	(50)	147	312	509
Moderate recycle	(50)	88	142	280
Plutonium balance (current plans)				
Maximum recycle	(+ 72)	+ 41	- 76	+ 37
Moderate recycle	(+ 72)	+ 100	+ 94	+ 266

With the exception of Italy, Spain and the UK, countries whose civil fuels are due to be reprocessed have plans to recycle plutonium in MOX fuels in their power reactors. The extent to which recycling will occur remains very uncertain (see chapter 7, sections V and VI). While MOX fabrication capacities are being expanded, the licensing of one of the main facilities (in Germany) is being opposed in the law courts. Furthermore, utilities are concerned about the high cost of MOX fuel, and in Germany and Japan they have yet to license sufficient reactors to take the fuel.

If Russia engages in plutonium recycling, it will have the option to use plutonium from its erstwhile military stockpile (see below). Leaving that aside, the maximum LWR–MOX fabrication capacity that could be installed by the end of the century is 360 tonnes of MOX fuel per year.[9] Once in place, these plants could have an annual consumption of around 18 tonnes of plutonium per year, assuming that the fuel is enriched in plutonium to a level of 5 per cent. Taking the optimistic view that a capacity of 300 tonnes per year would be reached by 1995, a further 130 tonnes of plutonium could be fabricated into MOX by the year 2000, or around 60 per cent of the envisaged stock of separated plutonium (see table 7.6). Under this maximum scenario, plutonium recycling would begin to eat into the plutonium surpluses after the turn of the century.

Under scenarios which do not envisage a reduction in the rate of reprocessing, and take a less bullish view of recycling, substantial surpluses of plutonium would continue to grow. Under the 'moderate' consumption scenario presented in chapter 7, for instance, plutonium surpluses would increase from 72 tonnes

---

[9] This would comprise 120 tonnes of heavy metal per year at both Marcoule in France and Hanau in Germany, 70 t/y. at Dessel in Belgium, and 50 t/y. at Sellafield in the UK.

in 1990 to 172 tonnes in 2000 and 266 tonnes in 2010. Under the 'stretch-out' scenario for plutonium production set out in chapter 6 and this 'moderate' consumption scenario, demand would be brought into a closer balance with supply, although there would still be a considerable excess (see table 7.7). If large plutonium surpluses are to be avoided, countries intent on reprocessing some of their fuel will have to renegotiate their contracts to bring supply into line with genuine demand.

Plutonium and HEU released from dismantled warheads

Before the breakup of the Soviet Union, the US and Soviet governments had already begun to instigate substantial arms reductions. The INF Treaty had eliminated stocks of intermediate-range nuclear weapons, and the START Treaty looked forward to significant reductions in strategic weapons. The disarmament process is now accelerating.[10] Presidents Bush and Gorbachev announced additional reductions in September and October 1991. Accompanying the formal confirmations by Belarus, Kazakhstan and Ukraine that they would eliminate all nuclear weapons on their territories, Presidents Bush and Yeltsin made commitments at the Lisbon summit meeting in May 1992 to reduce further the US and Russian arsenals.

It is not possible in 1992 to predict how far these reductions will go, and how quickly they will be implemented. Nor is it possible to assess how soon plutonium and HEU will be extracted from warheads. The 'de-construction' of warheads involves three stages, none of which is straightforward: the gathering together and storage of warheads at dismantling sites; dismantling the warheads and storing the components; and the processing of fissile components, bringing the plutonium and HEU to a form that is suitable for storage, recycling or disposal. As the CIS and the USA gear up to undertake these tasks, it is apparent that it will take many years to dismantle even a proportion of their arsenals. Present indications are that the CIS and the USA may not go beyond the second stage for a number of years.

Given these problems and uncertainties, tables 12.10 and 12.11 are limited to illustrating the amounts of material that could become available at different levels of arms reduction. The tables assume that the USA and the CIS today have 19 000 and 32 000 weapons respectively (the latter figure including old warheads now in storage); that China, France and the UK together have 1200 warheads;[11] and that each warhead contains 3.5 kg of plutonium (the mean of 3–4 kg) and 15 kg of HEU on average. The exception to this last rule is China, for which an average warhead content of 2.5 kg of plutonium and 20 kg of HEU is assumed—an average that has also been applied to the estimated 5000

[10] For a discussion of the problems and opportunities posed by the breakup of the former Soviet Union and its nuclear weapon arsenal, see Walker, W., 'Nuclear weapons and the former Soviet republics', *International Affairs*, vol. 68, no. 2 (Apr. 1992), pp. 255–77.
[11] This assumes that China has 400, France 500 and the UK 300 warheads.

Table 12.10. Illustrative amounts of weapon-grade plutonium released from dismantled nuclear weapons

Figures are in tonnes.

Number of warheads per arsenal	USA		Russia		UK, France and China	
	A	B	A	B	A	B
30 000	..	30	5	23	..	7
20 000	..	30	37	55	..	7
15 000	14	44	54	72	..	7
10 000	32	62	72	90	..	7
5 000	49	79	89	97	..	7
–	67	97	107	125	4	11

A: Weapon-grade plutonium that would be released from warheads after reducing to the warhead numbers in the left-hand column.
B: A plus weapon-grade plutonium already held outside warheads.

Table 12.11. Illustrative amounts of weapon-grade uranium released from dismantled nuclear weapons

Figures are in tonnes.

Number of warheads per arsenal	USA		Russia		UK, France and China	
	A	B	A	B	A	B
30 000	..	265	40	255	..	20
20 000	..	265	205	420	..	20
15 000	60	325	280	495	..	20
10 000	135	400	355	570	..	20
5 000	210	475	430	645	..	20
–	285	550	505	720	20	40

A: Weapon-grade uranium that would be released from warheads after reducing to the warhead numbers in the left-hand column.
B: A plus weapon-grade uranium already held outside warheads.

old warheads held in store in the former USSR. Figures in the tables are based on the central estimates of plutonium and weapon-grade uranium inventories derived in chapters 3 and 4. The quantities are rounded down to the nearest tonne.

Three points pertaining to tables 12.10 and 12.11 deserve emphasis:

1. The quantities of weapon-grade plutonium and HEU existing outside weapons are already very large. In total, there could be as much as 60 tonnes of plutonium and 540 tonnes of HEU in store, in old warheads that have not been dismantled, in scraps and in process residues. The exact composition of this

inventory outside weapons is extremely ill-defined. Even without warhead dismantlement, these inventories would have to be accounted for and dealt with.

2. Plutonium from dismantled weapons will, of course, mainly accumulate in the USA and Russia (assuming that the CIS warheads will be dismantled in Russia). Being optimistic, if each country dismantled all but 5000 of its warheads over the next decade, some 140 tonnes of weapon-grade plutonium and 640 tonnes of weapon-grade uranium would be released. Adding the stocks of military material held outside weapons, the quantities potentially available would be around 200 and 1180 tonnes respectively. The amounts of HEU are therefore very much larger than the amounts of plutonium. Also taking into account the greater ease with which the diluted HEU can be used in commercial reactors, it is apparent that these stocks of HEU will have a much greater impact on nuclear fuel markets than the plutonium.

The amounts of plutonium that could be derived from dismantled British, Chinese and French warheads, should those countries also decide to disarm, would be two orders of magnitude less, reflecting the much smaller sizes of their weapon arsenals. However, it should be recalled from the discussion in chapter 6 that rather similar quantities of plutonium (more than 200 tonnes over the next decade) are expected to arise from the expansion of civil reprocessing in Britain and France.

3. No strategies have yet been developed for dealing with these very large stocks of weapon-grade plutonium and uranium. This issue is covered in chapter 13.

13. Two policy issues

During the preparation of this volume, we have found ourselves returning time and again to two problems. The first is the continuing paucity of information about plutonium and HEU stocks in a number of vital areas. The second is the approaching surfeit of separated materials, as spent fuels are reprocessed and warheads are dismantled. These problems give rise to two important and related policy questions. What obligations should be placed on *all* states to reveal information about their inventories of plutonium and HEU? And what should happen to the hundreds of tonnes of plutonium and HEU that are about to be released from warheads and spent fuels? This study concludes with some brief comments on these two issues.

I. The need for an international register of plutonium and HEU

This book goes further than any previous one in setting out what is known about the world's inventories of plutonium and HEU. Despite detailed investigation, it is apparent that serious gaps remain in both public and private knowledge of these inventories. Error margins that can be as high as plus or minus 30 per cent, on stocks extending to hundreds of tonnes of material, speak for themselves.

Secrecy has been inherent to nuclear activity since its earliest days. To an unusual degree among military doctrines, nuclear deterrence was a game of deception, particularly at the height of the cold war in the 1950s and 1960s. Access to information, and the ways in which it was used, also became matters of great sensitivity in nuclear commerce, particularly when the nuclear industry became vulnerable to public opposition. However, the secrecy was double-edged: in the international context, it fed mistrust and became a dangerous source of instability.

One of the fundamental purposes of nuclear arms control was, therefore, to build confidence in intentions through specific disclosures of information. From the 1960s onwards, the USA and the USSR exchanged increasing amounts of information about their nuclear arsenals, at the same time as they subjected each other to intense scrutiny through intelligence operations. Furthermore, steps were taken to create an international framework within which countries without nuclear weapons could volunteer information about their nuclear activities—information which could then be verified by specially appointed international safeguards agencies.

The requirements and procedures for divulging and verifying information about plutonium and enriched uranium were formalized by the establishment of the European Atomic Energy Community (Euratom) in 1957, by the 1968 Treaty on the Non-Proliferation of Nuclear Weapons (NPT), and in the associated safeguards documents that were issued by Euratom and the IAEA.[1] They brought into being a regime that allowed for partial, qualified transparency. The transparency was *partial* in the NPT context because NWS parties were not required to divulge information on *any* of their nuclear activities, and because a number of countries outside the NPT refused to submit materials and facilities to international safeguards. In the Euratom context, the UK and France had to report on civil activities, but not on military activities. The transparency was *qualified* because information was imparted to the safeguards agencies in confidence. It remained private rather than public information. Furthermore, the needs of verification had to be balanced against the needs of the operator.

As is made clear in chapter 12, a consequence of the partial transparency required by the NPT is that the majority of the world's stocks of plutonium and HEU are today subject to no international supervision. The NPT regime is blind to them. We have estimated that some 30 per cent of the world stock of plutonium is safeguarded by the IAEA. It safeguards less than 1 per cent of the stock of HEU. The main reason is, of course, that the huge quantities of military and civil material held by the USA and the former Soviet Union are neither reported to the Agency nor submitted to international inspection.

The traditional justification for exempting the NWS inventories from IAEA safeguards has been that the IAEA's purpose is to detect the diversion of material from civil to military purposes, and thereby to discourage nuclear proliferation. As such, safeguarding in the NWS was only deemed worthwhile if it allowed the Agency to acquire technical expertise that could be applied in the NNWS. This was the main criterion guiding the Agency's selection of facilities for inspection under the 'voluntary offer' agreements it concluded with the NWS.

This position seems increasingly questionable, for three main reasons. The first is that the above arrangements did not envisage the breakup of a major nuclear weapon state, and the weakening of central control over nuclear stockpiles that it might entail. The fate of the hundreds of tonnes of weapon material in the former Soviet Union, not to mention the fate of the weapons themselves, is a matter of justifiable international concern. The diversion of materials from their intended use can no longer be regarded as an issue affecting the NNWS alone. Moreover, diversion can no longer be defined just in terms of movements from civil to military activity—we now have to guard increasingly against movements from one military purpose to another military (or paramilitary) purpose, within and across former boundaries. As such, military material has lost some of its former sanctity and its right to be kept outside the purview of the international community.

---

[1] The Tlatelolco Treaty, which covers Latin America, transfers safeguards responsibilities to the IAEA. Unlike Euratom, it does not establish a regional safeguarding system.

The second reason is that we have now entered an era of substantial, if not yet total, nuclear disarmament. While secrecy seems intrinsic to the processes of nuclear armament, disarmament allows and requires openness. Now that East and West are no longer threatening each other with nuclear obliteration, the strategic need to disguise their activities is much diminished. Furthermore, disarmament will involve the 'de-construction' of both weapons and large parts of weapon production systems, and the movement of materials and institutions towards the civil domain. The weapon-grade materials coming out of warheads will have to be properly supervised and accounted for in order to ensure that they are not being held back or spirited away, to ensure discipline and efficiency in what will clearly be a difficult and sophisticated industrial activity, and to monitor their re-entry into civil commerce where that occurs. There is a legitimate public interest in knowing how all these activities are being carried out, and in understanding how they will impinge on civil nuclear policy.

The third reason is the traditional concern for equity. The NPT is an inequitable Treaty that contains within it the aspiration to become equitable. It recognizes two distinct categories of country—nuclear and non-nuclear weapon states—but enjoins the former to take steps towards disarmament and thus towards entering the ranks of the latter. The radically different standards applied to the disclosure of information for these categories of countries are central to the NPT's inequity, and will have to change if its aspirations are to be even partially satisfied.

The threshold states provide the other context in which the lack of transparency is a matter of grave concern. Chapters 9, 10 and 11 show that the amounts of plutonium and HEU in those states are, by comparison, small—amounting to less than a tonne of either material—and that significant doubts remain about the effectiveness of the production capabilities that these countries have acquired. Nevertheless the clandestine activities of Iraq and North Korea have recently demonstrated how threatening nuclear weapon programmes can become even when their abilities to produce significant quantities of material are far from clear. The Persian Gulf War has been followed by the UN's unprecedented action compelling Iraq to disclose information about its nuclear weapon programme, to dismantle its capabilities, and to hand over whatever nuclear materials it has acquired to the custody of the IAEA. The UN Security Council's actions seem to have ushered in a more pugnacious approach to non-proliferation policy, and one that may be less respectful of the confidentiality that has marked the traditional relationship between governments, operators and safeguards agencies. Nuclear diplomacy may be becoming more open and confrontational.

How the UN's interventions in Iraq will affect the general conduct of nuclear relations remains the subject of much debate. Whatever the outcome, the international community seems to have arrived at a point at which there is a need and an opportunity to establish new international norms for information disclosure which would apply to the NWS and threshold states as well as the NNWS.

The norms established in the 1960s and early 1970s now seem in need of revision.

As one step in this direction, we believe that an international register of plutonium and HEU stocks and arisings is now required. Whether it is the IAEA, the UN or another accredited organization, some international body should be entrusted with the task of collecting and publishing annual data on plutonium and HEU. In parallel, governments should volunteer to publish details of the inventories existing on their territories at the end of each year, and to submit this information to the chosen international authority. This goes for nuclear as well as non-nuclear weapon states. (The UK has already set a good example in annually publishing civil plutonium figures.)

This exercise need not require such precise reporting as occurs in the application of safeguards, which it would in no way replace. Its main aim would be to provide policy makers with a reliable picture of the sizes and locations of plutonium and HEU stocks, and of the main changes occurring to them. Our experience is that most policy makers frequently have to make judgements in a state of near blindness. They only know, and are only allowed to know, what is happening in their own bailiwicks. Even there, they can be surprisingly illinformed. With so many important decisions approaching on how to deal with these materials, this no longer seems acceptable.

With regard to individual NNWS, the register should contain the following information at minimum (annual and cumulative amounts would be reported):

1. Estimates of the quantities of plutonium and HEU in spent fuels, and as yet undischarged from reactors;
2. The amounts of HEU and separated plutonium held in fresh and irradiated form;
3. The amounts of plutonium and HEU belonging to the country that are held abroad, and the amounts that are held in spent fuel or in unirradiated separated form;
4. Information on the facilities active in the production of HEU, in the separation of plutonium from spent fuels, and in the fabrication of fuel elements containing these materials.

The NWS should submit the same information on their civil programmes. The key question is how much they might be expected to divulge about their military programmes. In our view, they could provide total estimates of the amounts of plutonium and HEU assigned to nuclear weapons and nuclear-powered submarines without in any way endangering their security or imparting new information about nuclear weapon designs. At a minimum, they could provide the following information to the register:

1. The amounts of plutonium and HEU assigned to nuclear weapon arsenals;
2. The amounts of plutonium and HEU in military inventories that are not contained in warheads, and that play no active part in weapon production programmes or in submarine fuelling;

3. The amounts of plutonium and HEU that are extracted from dismantled weapons, and that are then held apart from or transferred to civil inventories;

4. The current status of production reactors and enrichment plants involved in the supply of plutonium and HEU for weapon programmes.

Regarding the accuracy requirements on the information imparted by both NWS and NNWS, we would suggest a target error margin of plus or minus 5 per cent. If individual governments can do better than this, all well and good.

II. The management of plutonium and HEU surpluses

Very large amounts of plutonium and HEU will be released from weapons and spent reactor fuels in the next two decades. If current reprocessing plans are implemented, around 215 tonnes of plutonium will be separated between the time of writing and the year 2000, and a further 235 tonnes in the following decade. If weapon dismantlement proceeds as expected, another 150 tonnes of plutonium and 500 tonnes of weapon-grade uranium could be released. These quantities will be added to today's already substantial stocks of these materials.

At present, there are no clear strategies for managing these unprecedented flows of material. This is not surprising since they were not anticipated, at least in the context of weapon dismantlement. Six broad observations can be made on the commercial prospects for using the materials as fuels in civil power reactors, and on the problems of control that will have to be addressed.

First, the HEU extracted from nuclear weapons and released from the large strategic reserves held by the USA and Russia will have the greatest impact on nuclear fuel markets. Its dilution into low-enriched uranium for power reactors is technically straightforward, while the recycling of plutonium is problematic; and the size and thus the fissile content of the HEU stocks are much greater than those of plutonium. Whereas four or five times as much HEU as plutonium is used in nuclear warheads in the USA and the CIS, the fission energy that can be extracted from a kilogram of each material is roughly the same in a thermal reactor.

The HEU in warheads due to be dismantled will translate into about two years' supply of enriched uranium fuel for the world's light water reactors, and will raise billions of dollars for the governments selling it (money that could help meet the costs of weapon dismantlement). The effects on fuel prices, and on demand for natural and depleted uranium, will depend on many factors, not least of which will be the rate at which the HEU is delivered to the market. The expectation is that this 'windfall' of enriched uranium will prevent fuel prices rising significantly above today's low levels for many years to come. Besides the public interest in more information, there is therefore also a strong commercial desire for data.

Second, while there is a ready market for HEU, it still presents security problems. The extraction of HEU (and plutonium) from warheads, its conversion from weapon- to reactor-grade material, its transportation between sites and between Russia and the USA under the agreement concluded in 1992 (whereby the latter will purchase the majority of the former's HEU stock)—at all stages, strict monitoring and physical protection will be required to prevent HEU being stolen or diverted.

The market is unlikely to be able or willing to absorb such large amounts of enriched uranium in just a few years. It nevertheless seems important that the dilution of HEU to low-enriched uranium should proceed ahead of consumption, so that the quantities of weapon-grade material are quickly reduced. Equally, all HEU stocks released from military programmes, and in the process of dilution, should be placed under IAEA safeguards.

Third, the reduction of plutonium stocks will be much more difficult. The radiological hazards associated with plutonium increase substantially the costs of fabricating the mixed-oxide (MOX) fuel elements containing plutonium. Furthermore, the use of plutonium involves some sacrifice in fuel efficiency in power reactors since higher burn-ups can be achieved with conventional enriched uranium fuels. The prospect of so much diluted HEU entering the market will make the burning of plutonium even harder to justify on grounds of either commercial advantage or supply security. If current reprocessing schedules are followed, the scale of MOX fabrication capacity in Europe and Japan is also insufficient to prevent a substantial proportion of the plutonium emanating from reprocessing plants being left in store.

While the material extracted from nuclear weapons will contain less of the radiologically troublesome isotopes of plutonium and will thus be easier to handle, the same disincentives will apply. In addition, neither Russia nor the USA have much commercial experience of plutonium recycling, and in Russia's case there will be particular worries that recycling might aggravate the safety problems already bedevilling its nuclear power programme.

Although warhead dismantlement will take many years to achieve, the benefits of disarmament are likely to outweigh by far the costs of dealing with the plutonium surpluses. In the commercial context, however, most utilities around the world now see more costs than benefits in separating plutonium from spent fuel. This being the case, questions about the validity of persevering with reprocessing in the face of such large plutonium surpluses are likely to become more insistent. A thorough re-examination of spent fuel policies in the areas where reprocessing is still practised (notably Western Europe, Japan and Russia) is already overdue.[2]

Fourth, it follows that if plutonium arisings cannot be absorbed commercially, much of the plutonium will have to be treated as waste. Hitherto, the assumption guiding many R&D programmes in the nuclear field has been that plutonium is an asset. As a result, next to nothing has been spent on developing techniques for getting rid of plutonium once it has been separated. Various sug-

---

[2] See, for instance, Albright, D. and Feiveson, H. A., 'Plutonium recycling and the problem of nuclear proliferation', *Annual Review of Energy,* vol. 13 (1988), pp. 239–65; and Berkhout, F., Suzuki, T. and Walker, W., 'The approaching plutonium surplus: a Japanese/European predicament', *International Affairs,* vol. 66, no. 3 (1990), pp. 525–43.

gestions have recently been made, including burning up the plutonium in specially designed reactors, sequestering it in rock formations by carrying out underground nuclear explosions, mixing it with high-level wastes, and dispatching it to the sun. None of these options have yet received rigorous development and testing to find out whether they are advisable on technical, economic and environmental grounds. A substantial international R&D effort is therefore going to be required to find acceptable solutions other than storage.

Fifth, while the control of HEU stocks will not be straightforward, plutonium presents special problems. The main stocks of HEU are the property of just two countries, the USA and Russia.[3] In contrast, six countries (the USA, Russia, the UK, France, Germany and Japan) will each possess tens of tonnes of separated plutonium, and a number of others will acquire smaller stocks. Many more countries, including both NWS and NNWS, are therefore likely to acquire large stocks of separated plutonium.

In addition, plutonium may have to be held in store for considerably longer periods than HEU if large-scale recycling cannot be managed and if, as seems likely, it takes many years to devise and implement plans for disposing of waste plutonium. It should also be recalled that plutonium is an unusually hazardous material so that its handling requires exceptional care.

For all these reasons, the proposals mooted in the 1970s that an international plutonium storage (IPS) scheme be established need to be looked at afresh.[4] The possibility that plutonium (and HEU) could be brought under IAEA lock and key is also anticipated in Article XII.A.5 of the IAEA Statute. At the very least, all separated plutonium outside weapons and the weapon production system should automatically be placed under IAEA safeguards, wherever it is located.

Finally, this discussion is mainly concerned with the need to guard against the potentially destablizing effects of these very large quantities of HEU and plutonium, just as disarmament is gathering momentum. But it is also important to recognize that the growth of these stocks will occur in a period when utilities and governments are facing major decisions about the future of nuclear power.

The present stock of power reactors was largely installed in the 1960s and 1970s, and will approach the end of its operating life in the 1990s and in the decade after the turn of the century. The highest priority for utilities will be to establish the political and economic conditions in which a new phase of reactor construction can be launched. They will, therefore, have strong interests in limiting their exposure to activities, such as plutonium recycling, that increase the costs and risks associated with nuclear power, and they will be as keen as anyone to ensure that plutonium and HEU are subject to stringent international control. Commercial and security interests should therefore coincide.

---

[3] However, it remains to be seen whether the Ukraine and Kazakhstan will claim title to HEU recovered from weapons once located on their territories, or how they might be reimbursed should Russia decide to sell it.

[4] Scheinman, L. and Fischer, D., 'Managing the coming glut of nuclear weapon materials, *Arms Control Today*, vol. 22, no. 2 (Mar. 1992).

Appendices

Appendix A. Enrichment technologies

Enriched uranium is produced in uranium isotope separation facilities. These facilities enrich the fraction of ^{235}U relative to ^{238}U, and are therefore called enrichment plants. Since the isotopes of an element have very similar chemical and physical properties, their separation is especially difficult. In an effort to reduce costs, countries have developed several different types of enrichment technology. Table A.1 describes the most common types. Table A.2 shows the amount of weapon-grade uranium, defined here as uranium containing more than 93% of the isotope ^{235}U, that can be produced given different assumptions about separative work units (SWU), the enrichment of uranium feed and the proportion of ^{235}U left in the tails (the tails assay).

Table A.1. Overview of principal enrichment technologies[a]

Technology	Status[b]	Countries involved[c]	Elementary separation factor[d]	Energy use (kWh/SWU)[e]
Gaseous diffusion	Major industrial use	USA, Russia, UK, France, China, Argentina	1.004	2 500
Gas centrifuge	Major industrial use	Russia, UK, France, Germany, Holland Japan, Pakistan, Brazil, Iraq, India, China	1.5	135
Separation nozzle	Large pilot plant built	Brazil	1.015	3 600
UCOR process	Commercial-size plant	South Africa	1.030	4 000
Chemical (CHEMEX process)	Pilot plant built	France	1.002	600
Chemical exchange (ASAHI process)	Pilot plant operating	Japan	1.001	150
Atomic Vapor Laser Isotope Separation (AVLIS)	Engineering development research only	USA, France, UK, Japan, Israel, Brazil	High	c. 100–200
Molecular laser (MLIS)	Development	Japan, Germany, South Africa	High	c. 235
Plasma separation (PSP)	Development	France	High	c. 330

[a] The primary source of this table is *Fiscal Year 1989 Arms Control Impact Statements,* Committees on Foreign Affairs and Foreign Relations of the House of Representatives and Senate, 100th Congress, 2nd session (US Government Printing Office: Washington, DC, Apr. 1988), p. 142.

[b] This column on the status of the technology refers only to the largest programmes.

[c] Russia and China have research and development programmes in various enrichment technologies that are not listed in this table.

[d] The elementary separation factor is defined as the ratio of the relative amount of ^{235}U in the product stream to the amount in the waste stream.

[e] The specific energy requirement, usually given in units of kilowatt-hour per separative work unit (kWh/SWU), is given for comparative purposes only. Actual values can vary.

Table A.2. Weapon-grade uranium production (93% enriched)

Capacity (SWU/y.)	Uranium feed (% ^{235}U)	Uranium^a tails (% ^{235}U)	WGU^b (kg/y.)	No. of days to produce 25 kg
5 000	0.71	0.2	21	435
	0.71	0.3	24	380
	0.71	0.5	31	295
5 000	5.0	1.0	122	75
	5.0	2.0	156	60
	5.0	4.0	208	45
15 000	0.71	0.2	63	145
	0.71	0.3	72	125
	5.0	1.0	366	25
70 000	0.71	0.2	294	31
	5.0	1.0	1 700	5

^{a} As the tails assay is raised, the amount of uranium feed required rises dramatically. Assuming that the uranium feed is natural uranium (0.71% ^{235}U), the production of 25 kg of weapon-grade uranium requires 4600 kg of natural uranium at a 0.2% tails assay, and about 11 000 kg of natural uranium at a 0.5% tails assay.

^{b} The following formulae relating separative work units (D), feed (F), tails (T), product (P), feed assay (N_F), tails assay (N_T) and product enrichment (N_P) are from Krass, A. S., Boskma, P., Elzen, B. and Smit, W. A., *Uranium Enrichment and Nuclear Weapon Proliferation*, SIPRI (Taylor and Francis: London, 1983), pp. 97, 99 and 100.

$$F/P = (N_P - N_T)/(N_F - N_T)$$

$$T/P = (N_P - N_F)/(N_F - N_T)$$

$$D/P = V(N_P) + V(N_T)[T/P] - V(N_F)[F/P]$$

$$V(N) = (2N-1)\ln[N/(1-N)]$$

Appendix B. Calculation of plutonium production in power reactors

The method used to derive estimates of plutonium production in power reactors in chapter 5 follows closely the method used by the Systems Studies Section in the Department of Safeguards of the International Atomic Energy Agency (IAEA) in studies published during the 1980s.[1] This method models the discharge of spent fuel from nuclear reactors, together with an estimate of their burn-up. Knowledge of mean fuel burn-up allows estimation of plutonium production in the fuel using standard conversion factors. These factors are shown as graphs in chapter 5—figures 5.3 and 5.4. In this appendix, only the method of calculation is explained. In the interests of brevity, the spreadsheets detailing electricity, fuel and plutonium production at each of the world's operating, decommissioned and planned reactors are not included. All the main results are discussed in chapter 5.

For each operating reactor, an estimate of heat output was calculated using published electricity generation data over four time periods: to the end of 1980; to the end of 1990; to the end of 2000; and to the end of 2010. Using published data or assumptions about reactor fuelling, mean fuel burn-up was estimated by reactor for each of these time periods. Plutonium production was estimated by employing conversion functions which relate mean burn-up in nuclear fuel to the generation of plutonium.

A simpler approximation for estimating plutonium in spent fuel that was used by one of the authors in earlier work.[2] In his previous work Albright used a conversion factor which matched electrical output with plutonium production by reactor *type*. For example, for LWRs, 330 kg of plutonium are estimated to be produced per GWe (net)-year. In general, each reactor of a particular type was assumed to have equivalent characteristics.

[1] For example, Bilyk, A., *Forecast of Amounts of Plutonium at Power Reactors subject to Safeguards (1981–1990)*, STR-125 (IAEA Department of Safeguards: Vienna, June 1982); and Mal'ko, M., *Estimation of Plutonium Production in Light Water Reactors*, STR-226 (IAEA Department of Safeguards: Vienna, Nov. 1986).

[2] Albright, D., *World Inventories of Plutonium*, PU/CEES Report no. 195 (Center for Energy and Environmental Studies: Princeton University, Princeton, N.J., June 1987).

Table B.1. Plutonium discharge rate by reactor type

Discharge is measured at 100% capacity factor.

Reactor type	Pu discharged[a] kg Pu per GWe (net)	Pu discharged[a] kg Pu per GWe (gross)
PWR	325	315
BWR	345	330
CANDU	630	570
GCR	815	735

[a] We assume mean burn-ups of PWR, 30.4 GWd/t; BWR, 28 GWd/t; CANDU, 7.5 GWd/t; and GCR, 4 GWd/t. Gross and net thermal efficiencies are typical values for these types of reactor (PWR [net: 0.320; gross: 0.332]; BWR [net: 0.312; gross: 0.328]; CANDU [net: 0.29; gross: 0.32]; GCR [net: 0.28; gross: 0.30]). The quantity of fuel discharged by each reactor type could then be calculated as follows:

Spent fuel = (365 * 1/n)/ burn-up,

where 365 is the electricity production in GWd from a 1-GW reactor operating at 100% capacity for one year, n is the thermal efficiency, and burn-up is expressed in GWd/t.

Finally, this fuel quantity was multiplied by the appropriate plutonium concentration by weight of fuel shown in figures 5.3 and 5.4.

The values in table B.1 closely match the ones in the earlier study, but there are important differences in the approach used here. In particular, an attempt has been made to use all available information on fuelling, achieved fuel burn-ups and actual thermal efficiencies of reactors. The overall effect is to lower estimates of plutonium discharges.

There are a number of advantages to the more sophisticated method used here. First, for all reactors, except those in the former Soviet Union, weight and average burn-up characteristics of fuel discharged by individual reactors are included in the models. This method does not assume that all reactors of a certain type always operate in the same way. Mean fuel burn-ups do fluctuate significantly from reactor to reactor, and this will have an effect on plutonium production, from reactor to reactor and within the lifetime of any single reactor. For instance, we estimate that mean burn-ups at British Magnox reactors in the decade 1981–90 varied between 3710 MWd/t (Hunterston 'A', total plutonium produced, 2.3 kg/t fuel) to 5080 MWd/t (Hinkley Point 'A', total plutonium produced, 2.86 kg/t fuel). While the difference between the fuel burn-up is 27%, plutonium production per kilogram of fuel differs by only 20%.[3] An attempt to encompass the non-linear relation between burn-up and plutonium production is therefore central to our analysis.

Second, this method takes account of the lower inventories of plutonium found in fuel discharged during the first few years of a reactor's operating life. For instance, at a PWR with a one-third core refuelling scheme, the first discharge of fuel after one year of operation will have a mean burn-up of one-third of equilibrium burn-up. In the

[3] The earlier method of plutonium calculation for these two reactors gives totals of 3.27 t Pu_{tot} for Hinkley Point 'A' (4.45 GWy.(e) (gross) generated in 1981–90) and 2.06 t Pu_{tot} for Hunterston 'A' (2.8 GWy.(e) (gross)). By comparison, the current method gives plutonium totals of 3.23 t Pu_{tot} for Hinkley Point 'A' and 2.09 t Pu_{tot} for Hunterston 'A'. Using values from table B.1, therefore, gives results which in this case agree closely with those in the present method.

second year, mean burn-up will be two-thirds of equilibrium, and only at the third discharge will the mean burn-up reach around 33 000 MWd/t. The effect on plutonium production is significant in the first decade in which a reactor starts-up or is decommissioned. Again, the main effect is to bring plutonium estimates down.[4]

Third, this method allows a more sophisticated perspective on how different fuel cycles work, and in particular generates spent fuel figures. These have some worth in themselves, but have the further advantage in that they can be corroborated with spent fuel figures published in the UK, the USA, Germany, Japan and France. This provides another means of corroborating plutonium estimates.

Fourth, it allows the generation of forecasts for plutonium production which can take account of trends in fuel management—higher fuel burn-ups, for instance. This will have the overall effect of bringing down fuel and plutonium discharges from reactors for a given electrical output.

This approach still contains many averaging assumptions. The initial enrichment of fuel and the neutron flux within a reactor core are not uniform. Fuel burn-up within the core, and within each individual fuel rod, therefore varies from place to place. Within any batch of fuel the burn-up of different fuel pellets may differ by as much as 20%, and will depend on the way the reactor has been operated and fuel within it has been moved around.

At the current level of data it would be impossible to embark on this type of analysis, and indeed the benefits of so doing are likely to be marginal.

Outline of method

A variety of approaches were used to calculate plutonium inventories, depending on the information available. In general, an assumption has been made that reactor operators have tried to achieve design burn-ups in order to reduce the costs of fresh fuel and the inconvenience of managing spent fuel. For further information on the different strategies used to derive plutonium figures, see chapter 5, section III.

For non-nuclear weapon states, we have used design equilibrium burn-up assumptions for each reactor published by the IAEA[5] to make estimates of the amount of fuel discharged using the cumulative gross electricity outputs published by *Nucleonics Week*. For reactors not under IAEA safeguards, such as those in the United Kingdom, France, the United States and India, we have estimated mean burn-ups using electricity outputs and assumptions about fuel discharges derived from published sources, or from fuel core characteristics published in the *World Nuclear Industry Handbook 1992*.[6] In many cases it has been possible to normalize these fuel discharge estimates by referring to published totals for each country.

---

[4] In the case of a PWR, the earlier method will overestimate cumulative plutonium discharges for the first decade by about 7–10%, assuming similar burn-ups to those in table B.1. For the first five years of operation, the previous method overestimates plutonium discharges by 15–25%. The reason is that burn-ups during the initial years are significantly less, lowering plutonium values sharply. During the second and third decades of operation, the two methods give similar results. For 20 to 30 years of cumulative discharges, the estimates differ by only a few per cent, again assuming similar burn-ups.

[5] Mal'ko, M., *Plutonium Production in Power Reactors subject to IAEA Safeguards (1985–1994)*, STR-228 (IAEA Department of Safeguards: Vienna, Dec. 1986), Annex: 'Input data for plutonium production forecasting'.

[6] *Nuclear Engineering International, World Nuclear Industry Handbook 1992* (Reed International: London, 1991).

Plutonium production at Soviet reactors, for which there are neither burn-up nor electrical output data, has been modelled using the assumptions about reactor fuelling and capacity factors (that is, the average proportion of the time at which reactors operate at full load). This is equivalent to the method used by Albright in earlier work..

The basic modelling steps for each of these three categories are as follows.

Power reactors in NNWS safeguarded by the IAEA and in France and the United States

Step 1: Calculate the amount of fuel discharged by the reactor in a given time period.

The total weight of irradiated fuel discharged is calculated using the function:

$$F_t = E_t / (24B [n/100]) - C \qquad (1)$$

where F_t is the total amount of fuel (tonnes) discharged in period t, E_t is gross electrical output of the reactor in time period t (GWh), B is mean equilibrium design fuel burn-up (GWd/t), n is the gross thermal efficiency of the reactor (%), and C is a half-core of fuel (t) added in the first and last time period in which a reactor operates (that is, if a reactor with a 30-year life began operating in 1978, C would be added in the time periods 'to 1980' and '2001–10').

The factor C is introduced to take account of the heat produced by partially irradiated fuel still held in the core (in the first time period) and by partially irradiated fuel discharged after the reactor has been decommissioned. Without this factor, estimates of fuel discharges would be over-estimates.

Example:

The Atucha 1 heavy water reactor in Argentina has a core of 38 t, a nominal design fuel burn-up of 5.6 GWd/t and a gross thermal efficiency of 30.3%. By the end of 1980 it had generated 15 689 GWh of electricity. We estimate that the amount of fuel discharged by Atucha 1 up to the end of 1980 is:

$$(15 689 / [24 \times 5.6 \times 0.303]) - 19 = 366 t$$

For weapon states such as France and the United States, whose reactors are not included in previous IAEA analyses, mean fuel burn-ups were estimated *by reactor type*, based on published figures for aggregate spent fuel discharges.

For example, US PWR discharges were estimated for each of the four time cohorts using a simple assumption of a one-third core reload per year. These estimates were compared with US Department of Energy, Energy Information Agency[7] estimates of the amounts of fuel discharged by US PWRs, which showed that our first estimates were too high. The fuel curve was adjusted iteratively by changing the fuel burn-up in equation 1 associated with each reactor until a 'best fit' was achieved. In the process, new figures were derived for the amount of fuel discharged by each reactor. The same procedure was carried out for BWR fuel estimates.

---

[7] See chapter 5, footnote 4.

Although this method does not provide much confidence in the accuracy of fuel and plutonium figures for individual reactors, it does allow the normalization of aggregate estimates of plutonium discharges data with available fuel discharge information.[8]

For United Kingdom reactors

Step 1: Calculate the mean burn-up of discharged spent fuel.

For the UK there exists a uniquely good base of information on reactor fuel discharges.

The mean fuel burn-up is estimated using the function:

$$B = E_t / (24n [F_t + C]).$$

Example:

The Oldbury Magnox station with a thermal efficiency of 28.8% generated 33 170 GWh of electricity in the period 1981 to 1990. During this period it discharged about 960 tonnes of fuel.[9] The mean burn-up of this fuel was therefore:

$$33\ 170 / (24 \times 0.288 \times 960) = 4.99\ \text{GWd/t of fuel.}$$

For all power reactors, excluding those in the former Soviet Union

Step 2: Calculate the mean plutonium content in the fuel.

Mean plutonium content in the fuel is derived using published or estimated mean fuel burn-up. This has been achieved using functions derived from the work of Bilyk and Mal'ko at the IAEA.[10] In deriving figures for plutonium production in natural uranium fuel at low burn-ups (less than 2000 MWd/t) we have used data produced by Turner.[11]

Apart from burn-up, the main determinant of the plutonium in irradiated fuel is the initial uranium-235 enrichment of the fuel. We have used five categories of enrichment: natural uranium (0.7% ^{235}U); advanced gas-cooled reactors (1.5%); boiling water reactors (2.7–2.8%); pressurized water reactors (3.2–3.3%); and VVERs (3.6%). We have also used these curves for calculations of plutonium generation at higher fuel burn-ups.

---

[8] The effect on plutonium estimates when compared with the simpler approximation in table B.1 varies from reactor to reactor. For example, in 1981–90 the Fort Calhoun 1 PWR generated 3.45 GWy(e), while the Peach Bottom 3 BWR generated 4.63 GWy(e). According to table B.1 these reactors would have discharged 1.09 t Pu_{tot} and 1.53 t Pu_{tot} respectively, compared with 1.0 t Pu_{tot} and 1.6 t Pu_{tot} in our estimates.

[9] This figure is derived from Barnham, K. W. J. *et al.*, 'The production and destination of British civil plutonium', *Nature*, vol. 317 (19 Sep. 1985), pp. 213–17, and from official declarations in *Parliamentary Debates, House of Commons Official Report [Hansard]*, 1 Apr. 1982, 27 July 1983, 25 Jan. 1985, 23 July 1985 and 21 July 1986, and in British Department of Energy Press Releases, 16 Dec. 1987, 13 Oct. 1988, 5 Dec. 1989, 18 Oct. 1990, and 17 Oct. 1991.

[10] Bilyk (note 1) and Mal'ko (note 1).

[11] Turner, S. E., *et al.*, *Criticality Studies of Graphite-Moderated Production Reactors*, SSA-125, Southern Sciences Applications, prepared for the US Arms Control and Disarmament Agency, Jan. 1980.

The functions used to calculate plutonium inventories in these fuel types are as follows:

Natural uranium fuel (Magnox and CANDU):

$$Pu_{tot} = 0.9235 \; B^{0.6946},$$

where Pu_{tot} is the mean plutonium content (kg/t of fuel).

This overestimates plutonium production at low burn-ups. The effect of using the Turner figures is to bring down plutonium estimates below 2000 MWd/t by up to 30%.

This function is shown graphically in figure 5.3.

Enriched uranium fuel

BWR:

$$Pu_{tot} = (1.138 \times 10^{-4}) \, B^3 \; - \; 0.011 \, B^2 \; + \; 0.508 \, B \; + \; 0.144$$

PWR:

$$Pu_{tot} = (0.978 \times 10^{-4}) \, B^3 \; - \; 0.011 \, B^2 \; + \; 0.523 \, B \; + \; 0.193$$

These two curves are shown in figure 5.4.

VVER:

$$Pu_{tot} = (0.551 \times 10^{-4}) \, B^3 \; - \; 0.008 \, B^2 \; + \; 0.480 \, B \; + \; 0.193$$

The curve for plutonium generation in AGR fuel shown in figure 5.4 was derived primarily from data published by the UK Department of Energy annual plutonium figures.[12]

For all power reactors, excluding those in the former Soviet Union

Step 3: Calculate amount of plutonium in discharged spent fuel.
    The total amount of plutonium discharged from each reactor is then estimated by multiplying mean plutonium content (Pu_{tot}) by the amount of fuel discharged (F_t).

Example:

The Biblis B PWR generated 76 846 GWh (gross) of electricity between 1981 and 1990. We estimate that it discharged 283 t of fuel at a mean discharge of 32.5 GWd/t.

---

[12] Department of Energy (note 9).

Mean plutonium content at this burn-up is 8.93 kg/t of fuel. The reactor therefore discharged a total of 2.53 t $Pu_{tot.}$[13]

For power reactors in the former Soviet Union

Steps 1 to 3: Estimate electricity production at each reactor and estimate plutonium production.

For reactors in Russia, Lithuania, the Ukraine and Armenia we have fallen back on the basic method used by Albright (1987). Plutonium production for these reactors is based on standard electricity-plutonium conversion factors for VVER and RBMK reactors. For VVERs we assume 330 kg Pu per GWe (net)-year, for RBMKs we assume 300 kg plutonium per GWe (net)-year.

We have estimated electricity generation by using standard capacity factors. We assume capacity factors of 75% for VVER 440s, and 65% for RBMKs and VVER 1000s.

Example:

The Beloyarsk 1 RBMK began operating in 1964 with a design rating of 102 MWe (net). Using the above assumptions we can estimate that by the end of 1980 the reactor had discharged,

$$(1980-1964) \times (0.102 \times 0.65) \times 300 = 320 \text{ kg of plutonium.}$$

Separate estimates of fuel discharge weights were made using nominal values for annual discharges published in the design reload values given in the *World Nuclear Industry Handbook* (note 6).

---

[13] By comparison, using the earlier factors yields a result of 2.76 t Pu_{tot}, an overestimate of about 8%.

Appendix C. Separation of plutonium from power reactor fuel at reprocessing plants

As with estimates of plutonium production in reactors, estimates of the amount of plutonium separated are composed of data derived from a wide variety of sources. In most cases these sources have already been given in footnotes in the main text. In this appendix historical data compiled by the authors are reproduced for six reprocessing plants: B205, UP1, UP2, UP3, WAK and Tokai-mura.

The main parameters in estimating plutonium separation are the throughput of fuel at a reprocessing plant, the fuel's burn-up at discharge, the length of time that the spent fuel was stored prior to reprocessing and the efficiency of the separation process. For many of the plants reasonably good fuel throughput figures are published or can be estimated from what is known about reactor fuel discharges. Burn-up information is more scarce, being available only for plants in France, Germany and the United States. In other cases we have made assumptions about burn-ups using a variety of sources, some published (as with the Tokai plant for which mean burn-ups over periods of years have been published) and some based on estimates made in our own calculations of plutonium production in power reactors. Details about the amount of plutonium lost in waste streams at reprocessing plants (process losses) are much harder to come by. We have assumed fixed values of 5% for older plants processing metallic fuel (B205 and UP1) and 1% for all other plants.

Table C.1. Plutonium separation from British Magnox power reactor fuel at B205[a]

Year	Fuel throughput (t)	Burn-up (MWd/t)	Plutonium separated (kg)	Cumulative Pu separated (kg)
1965	90	500	40	40
1966	140	1 000	110	150
1967	370	1 500	410	560
1968	470	1 500	520	1 080
1969	750	2 000	1 050	2 130
1970	1 000	2 000	1 395	3 525
1971	1 177	2 500	1 950	5 475
1972	1 086	2 500	1 800	7 275
1973	765	2 500	1 270	8 545
1974	730	3 000	1 375	9 920
1975	1 121	3 000	2 110	12 030
1976	589	3 500	1 235	13 265
1977	956	3 500	2 000	15 265
1978	867	4 000	1 995	17 260
1979	722	4 000	1 660	18 920
1980	858	4 200	2 040	20 960

(Table C.1 continued)

Year	Fuel throughput (t)	Burn-up (MWd/t)	Plutonium separated (kg)	Cumulative Pu separated (kg)
1981	888	4 200	2 110	23 070
1982	975	4 400	2 395	25 465
1983	1 024	4 300	2 475	27 940
1984	1 040	4 200	2 470	30 410
1985	781	3 100	1 505	31 915
1986	850	4 200	2 020	33 935
1987	861	3 500	1 805	35 740
1988	772	4 700	1 985	37 725
1989	787	4 900	2 080	39 805
1990	988	5 000	2 650	42 455
1991	688	5 000	1 845	44 300

[a] The assumptions used in making these estimates are described in section V in chapter 6. The burn-up figures used here are taken from mean burn-up calculations for British reactors. See appendix B for the methodology. We assume process losses of 5%.

Table C.2. Plutonium separated from foreign magnox fuel at B205[a]

Year	Fuel throughput (t)	Burn-up (MWd/t)	Plutonium separated (kg)	Cumulative Pu separated (kg)
1967	60	500	25	25
1968	100	1 000	80	105
1969	100	1 500	110	215
1970	100	2 000	140	355
1971	100	2 500	165	520
1972	100	3 000	190	710
1973	100	3 000	190	900
1974	100	3 000	190	1 090
1975	100	3 000	190	1 280
1976	100	3 000	190	1 470
1977	100	3 000	190	1 660
1978	100	3 000	190	1 850
1979	100	3 000	190	2 040
1980	100	3 000	190	2 230
1981	100	3 000	190	2 420
1982	100	3 000	190	2 610
1983	100	3 000	190	2 800
1984	130	3 000	245	3 045
1985	130	3 000	245	3 290
1986	130	3 000	245	3 535
1987	121	3 000	230	3 765
1988	32	3 000	60	3 825
1989	88	3 000	165	3 990
1990	141	2 500	235	4 225
1991	51	2 500	85	4 310

Note to table C.2

^{a} Estimates of throughputs of Japanese and Italian fuel at B205 before 1987 are based on fuel discharge estimates made in chapter 5. From 1987, the Department of Energy has published fuel and plutonium data (Department of Energy Press Releases, 16 Dec. 1987, 13 Oct. 1988, 5 Dec. 1989, 18 Oct. 1990, and 17 Oct. 1991). Burn-up assumptions are also based on estimates made in chapter 5.

Table C.3. Plutonium separated from Magnox power reactor fuel at UP1 (France)

Year	Fuel throughput (t)	Burn-up (MWd/t)	Plutonium separated (kg)	Cumulative Pu separated (kg)
1965	40	500	20	20
1966	49	500	20	40
1967–70	0	..	..	..
1971–72	3	2 000	5	45
1973	0	..	..	..
1974	111	2 000	160	205
1975	8	2 000	10	215
1976	21	2 000	30	245
1977	120	2 000	170	415
1978	245	2 000	350	765
1979	280	2 000	400	1 165
1980	267	2 000	380	1 545
1981	208	2 500	345	1 890
1982	300	2 500	495	2 385
1983	125	3 000	235	2 620
1984	315	3 000	595	3 215
1985	436	3 500	915	4 130
1986	387	3 500	810	4 940
1987	350	3 500	735	5 675
1988	327	3 500	685	6 360
1989	370	350	775	7 135
1990	330	3 500	690	7 825
1991	330	3 500	690	8 515

Sources: Syndicat CFDT de l'Energie atomique, *Le dossier électronucléaire* (Éditions du Seuil: Paris, 1981), pp. 186–91; Hirsch, H. and Schneider, M., *Wiederaufarbeitung in Europa: Wackersdorf ist tot—es lebe La Hague?*, Rest-Risiko, nr. 6, Greenpeace, Hamburg, Apr. 1990; Commissariat à l'Energie Atomique (CEA), 'Le retraitement des combustibles irradiés', *Industrie Nucléaire Française*, Paris, 1982, pp. 154–64; Couture, J., 'Status of the French reprocessing industry', paper given to the American Nuclear Society Conference, *Fuel Processing and Waste Management*, 26–29 Aug. 1984, Jackson, Wyo.; Delange, M., 'Operating Experience with Reprocessing Plants', *Atomwirtschaft*, Jan. 1985, pp. 24–28; Delange, M., 'LWR spent fuel reprocessing at La Hague: Ten years on', *RECOD 87* conference, Paris, 1987, pp. 187–93; 'Reprocessing and waste management, country: France, pt 1', *NUKEM Market Report*, no 3 (1988), pp. 15–18; Lewiner, C. and Gloaguen, A., 'The French reprocessing programme', *Atomwirtschaft*, May 1988, pp. 227–29; CEA, *Cycle du combustible nucléaire: retraitement*, Paris, Mar. 1989; 'Reprocessing and waste management: review 1989', *NUKEM Market Report*, no. 2 (1990), pp. 14–23; EdF, 'Retraitement recyclage', Paper by Service des Combustibles, Paris, 6 Mar. 1990.

Table C.4. Plutonium separated from Magnox power reactor fuel at UP2 (France)[a]

Year	Fuel throughput (t)	Burn-up (MWd/t)	Plutonium separated (kg)	Cumulative Pu separated (kg)
1966	52	500	25	25
1967	98	800	65	90
1968	189	1166	180	270
1969	157	986	130	400
1970	245	1 079	215	615
1971	126	2 287	205	820
1972	250	2 164	385	1 205
1973	213	2 385	355	1 560
1974	635	2 331	1 035	2 595
1975	443	3 038	870	3 465
1976	218	2 783	400	3 865
1977	355	2 947	680	4 545
1978	371	3 345	775	5 320
1979	240	3 590	530	5 850
1980	252	3 317	525	6 375
1981	250	3 672	560	6 935
1982	226	3 720	510	7 445
1983	117	3 272	240	7 685
1984	185	3 865	430	8 115
1985	120	3 900	280	8 395
1986	76	3 900	175	8 570
1987	77	3 900	180	8 750

[a] Magnox fuel was reprocessed at UP2 from 1966 to 1987. For sources, see table C.3.

Table C.5. Plutonium separated from oxide fuel at UP2 (France)[a]

Year	Fuel throughput (t)	Burn-up (MWd/t)	Plutonium separated (kg)	Cumulative Pu separated (kg)
1976	14.6	15 800	90	90
1977	18	28 000	150	240
1978	38	27 300	310	550
1979	79	20 400	560	1 110
1980	105	21 000	750	1 860
1981	101	25 400	800	2 660
1982	154	21 100	1 105	3 765
1983	221	23 200	1 670	5 435
1984	255	23 200	1 925	7 360
1985	343	24 000	2 635	9 995
1986	333	21 000	2 385	12 380
1987	425	23 500	3 250	15 630
1988	355	21 000	2 540	18 170
1989	430	21 000	3 080	21 250
1990	400	20 000	2 786	24 035
1991	400	20 000	2 785	26 820

[a] Throughput and fuel figures for 1989, 1990 and 1991 are estimated.

Table C.6. Plutonium separated from oxide fuel at WAK (Germany)[a]

Year	Fuel throughput (t)	Plutonium separated (kg)	Cumulative Pu separated (kg)
1971	3.45	14	14
1972	13.62	53	67
1973	7.26	21	88
1974	5.89	1	89
1975	11.80	53	142
1976	13.74	105	247
1977	15.76	127	374
1978	13.67	55	429
1979	19.17	81	510
1980	9.17	37	547
1981	..	..	547
1982	6.59	27	574
1983	16.81	110	684
1984	11.80	51	735
1985	11.68	105	840
1986	14.12	115	955
1987	14.31	103	1 058
1988	5.44	54	1 112
1989	3.04	25	1 137
1990	9.45	43	1 180

[a] These figures are actual outturns. Personal communication from Dr Lausch and Dr Zabel, WAK, Leopoldshaven, Karlsruhe, 20 Jan. 1992.

Table C.7. Plutonium separated from oxide fuel at Tokai-mura (Japan)

Year	Fuel throughput (t)	Burn-up (MWd/t)	Plutonium separated (kg)	Cumulative Pu separated (kg)
1977	3	10 000	10	10
1978	16	15 000	90	100
1979	5	17 000	30	130
1980	55	17 000	330	465
1981	41	18 000	255	720
1982	50	18 000	315	1 035
1983	3	19 000	20	1 050
1984	0	19 000	0	1 050
1985	79	21 000	545	1 600
1986	45	21 000	310	1 910
1987	50	23 000	360	2 270
1988	44	23 000	320	2 590
1989	18	23 000	130	2 720
1990	99	25 000	750	3 470
1991	90	25 000	685	4 155

Source: 'Tokai marks reprocessing of 500 tonnes of fuel', *PNC Review*, no. 17 (spring 1991), p. 6.

Appendix D. Research reactors (>1 MWth) using HEU fuel

These tables list the research and test reactors with a power rating greater than one megawatt-thermal that use HEU fuel. The listings for the former Soviet Union and China are derived from information provided by these countries to the IAEA. As a result, the listings might be incomplete. Where information is available, they indicate the current position regarding the conversion of research reactors to use low-enriched fuels.

Table D.1. US operating research and test reactors with power > 1 MWth using HEU fuel (as of end 1991)[a]

Reactor	Power (MWth)	Enrichment (%)	Kg ^{235}U/y.	Conversion
DOE reactors				
ATR	250	93	~175	Not planned
HFIR	100	93	~150	Not planned
HFBR	60	93	~64	Not planned
BMRR	3	90	~0.2	Not planned
BSR	2	93	~0.5	Not planned
Total	**415**		**~390**	
NRC-licensed reactors				
NBSR	20	93	8.7	?
MURR	10	93	19	Not planned
MITR	5	93	6.7	Not planned
GTRR	5	93	2	In process
RINSC	2	93	4	In process
UVAR	2	93	4	In process
ULR	1	93	1	In process
4 TRIGA reactors	4	70	0.6	In process
Total	**49**		**46**	

[a] Matos, J. E., *Estimated Uranium Densities with Reduced Enrichment for DOE Research and Test Reactors*, Argonne National Laboratory, 21 Oct. 1986.

[b] Excluded are fast reactors that use HEU fuels, in particular the EBR-2 which uses 67% enriched uranium fuel.

Table D.2. US-supplied operating research and test reactors with power > 1 MWth using HEU (>90%) fuel (as of May 1991)[a]

Reactor	Power (MWth)	Enrichment(%)	kg ^{235}U/y.	Conversion[b]
RA-3 (Argentina)	2.8	20	..	Yes
HIFAR (Austria)	10	75	7.5	
ASTRA (Austria)	8	20	..	Yes
BR-2 (Belgium)	80	93	27	Not agreed
IEA-R1 (Brazil)	2	20–93	1	
NRU (Canada)	125	93	65	
NRX (Canada)	24[c]	93	1	Shut in 1994
MNR (Canada)	5	93	1.9	
LO AGUIRRE (Chile)	10[d]	90	0	
LA REINA (Chile)	5	45–80	1	
DR-3 (Denmark)	10	20	..	Yes
OSIRIS (France)	70	7	..	Yes
RHF (France)	57	93	51	Not agreed
SILOE (France)	35	93	23	
SCARABEE (France)	20	?	0	
ORPHEE (France)	14	93	14.7	Not agreed
FRJ-2 (Germany)	23	80	18	
FRG-2 (Germany)	15	93	11	
BER-2 (Germany)	10	92	4.8	
FRG-1 (Germany)	5	20	..	Yes
FMRB (Germany)	1	93	1	
GRR-1 (Greece)	5	93	2.7	
NRCRR (Iran)	5	20	..	Yes
IRR-1 (Israel)	5	93[e]	0	?
JMTR (Japan)	50	45	35	
JRR-2 (Japan)	10	45	10	Shut in 1995
KUR (Japan)	5	93	2.1	
JRR-4 (Japan)	3.5	93	0.9	
TRIGA (Mexico)	1	70	0.7	
HFR Petten (Netherlands)	45	90	36	Not agreed
HOR (Netherlands)	2	20	1.7	1992?
PARR (Pakistan)	5	20	..	Yes
RPI (Portugal)	1	93	0.9	
PRR (Philippines)	1	20	..	Yes
SSR (Romania)	14	20–70	11	
SAFARI (S. Africa)	5	45	3	No plans
TRIGA (S. Korea)	2	20–70	1	
R-2 (Sweden)	50	20–90	27	
R2-0 (Sweden)	1	93	0[f]	
SAPHIR (Switzerland)	10	20–93	5.6	
THOR (Taiwan)	1	20	..	Yes
TR-2 (Turkey)	5	93	1.5	
HERALD (UK)	5	70	3	
Total	656[g]		370[g]	

Notes for table D.2.

[a] Letter from Armando Travelli, RERTR Program Manager, to Dr Leonard Weiss, Staff Director, US Senate Committee on Governmental Affairs, 19 Apr. 1988; and Travelli, A., 'The RERTR Program: a status report', paper presented at the 1991 International Meeting on Reduced Enrichment for Research and Test Reactors, 4–7 Nov. 1991, Jakarta, Indonesia.

[b] The absence of a comment in this column means that the reactor is in the process of conversion.

[c] In hot stand-by status.

[d] Operates sporadically.

[e] Out of HEU fuel, might not be operating.

[f] Functions as a critical facility.

[g] Total includes only reactors that have not been converted.

Table D.3. CIS operating research and test reactors with power > 1 MWth using HEU fuel (as of May 1991)[a]

Reactor[b]	Power (MWth)	Enrichment (%)	Kg ^{235}U/y.
IR-8	8	90	
IRT-A MEPI	2.5	90	
BR-10	8	90	
WWR-CM Tashkent	10	90	
WWR-M Kiev	10	36	
WWR-M Gatchina	18	90	
IRT-2	5	90	
IVV-2M Sverdlovsk	15	90	
MR	40	90	
MIR-M1	100	90	
IRT-T	6	90	
WWR-K Alma Ata	10	36	
WWR-TS	12	36	
RBT-10/2	10	90	
RBT-10/1	10	63	
RBT-6	6	63	
IRT-MIFI	10	?	
SM-2	100	90	
PIK physical model	?	90	
Total	**380**		250–350[c]

[a] IAEA, *Nuclear Research Reactors in the World* (IAEA: Vienna, 1991).

[b] Excluding fast reactors that use HEU fuels.

[c] Estimate based on 0.7 to 0.9 kg ^{235}U per MWth.

Table D.4 . CIS-supplied operating foreign research and test reactors with power
> 1 MWth using HEU fuel (as of May 1991)[a]

Reactor	Power (MWth)	Enrichment (%)	Kg ^{235}U/y.
LR-15 Prague (Czechoslovakia)	10	*80*	
WWR-SZM Budapest (Hungary)	5	*36*	
IRT-DPRK (N. Korea)	8	*80*	
IRT-1 (Libya)	10	*80*	
EWA (Poland)	10	*36*	
AGATA (Poland)	?	*80*	
MARIA (Poland)	30	*80*	
R-A (Yugoslavia)	6.5	*80*	
Total	**80**		**55–70**[b]
RFR (former GDR)	10	*36*	Closed, 1992?
? (Cuba, planned)	10?	*80?*	Reactor shipped

[a] IAEA, *Nuclear Research Reactors in the World* (IAEA: Vienna, 1991).
[b] Estimate based on 0.7 to 0.9 kg ^{235}U per MWth.

Table D.5. Chinese research and test reactors with power > 1 MWth using HEU fuel
(as of May 1991)[a]

Reactor	Power (MWth)	Enrichment (%)	kg ^{235}U/y.
HFETR	125	*90*	90–110[b]
PPR PULSING	1	*90*	1

[a] IAEA, *Nuclear Research Reactors in the World* (IAEA: Vienna, 1991).
[b] Estimate based on 0.7 to 0.9 kg ^{235}U per MWth.

Index